The Healthcare Library of Northern Ireland
and
Queen's University Biomed

Biomedical Li   y

BIOMEDICAL LIBRARY

Tel: 329241 Ext. 2797

j

i

THE MEDICAL PERSPECTIVES SERIES

Advisors:

David Hames *Department of Biochemistry and Molecular Biology, University of Leeds, U.K.*

David R. Harper *Department of Virology, Medical College of St Bartholomew's Hospital, London, U.K.*

Andrew P. Read *Department of Medical Genetics, University of Manchester, U.K.*

Robin Winter *Institute of Child Health, London, U.K.*

Oncogenes and Tumor Suppressor Genes
Cytokines
The Human Genome
Autoimmunity
Genetic Engineering
Asthma
DNA Fingerprinting
Molecular Virology
HIV and AIDS
Human Vaccines and Vaccination
Antibody Therapy
Antimicrobial Drug Action

Forthcoming titles:

Antiviral Therapy
Molecular Biology of Cancer

Antimicrobial Drug Action

R.A.D. Williams

Department of Biochemistry, Medical Sciences Building, Queen Mary and Westfield College, Mile End Road, London E1 4NS, U.K.

P.A. Lambert

Department of Pharmaceutical and Biological Sciences, Aston University, Aston Circle, Birmingham B4 7ET, U.K.

P. Singleton

1 Church Cottage, Clannaborough, Crediton, Devon EX17 6DA, U.K.

βIOS
SCIENTIFIC
PUBLISHERS

© **BIOS Scientific Publishers Limited, 1996**

First published 1996

A CIP catalogue record for this book is available from the British Library.

ISBN 1 872748 81 3

BIOS Scientific Publishers Ltd
9 Newtec Place, Magdalen Road, Oxford OX4 1RE, UK
Tel. +44 (0)1865 726286. Fax. +44 (0)1865 246823
World Wide Web home page: http://www.Bookshop.co.uk/BIOS/

DISTRIBUTORS

Australia and New Zealand
　　DA Information Services
　　648 Whitehorse Road, Mitcham
　　Victoria 3132

India
　　Viva Books Private Limited
　　4325/3 Ansari Road, Daryaganj
　　New Delhi 110 002

Singapore and South East Asia
　　Toppan Company (S)PTE Ltd
　　38 Liu Fang Road, Jurong
　　Singapore 2262

USA and Canada
　　BIOS Scientific Publishers
　　PO Box 605, Herndon, VA 22070

Typeset by Florencetype Ltd, Stoodleigh, Devon, U.K.
Printed by Information Press Ltd, Oxford, U.K.

Contents

Abbreviations

ABC	ATP-binding cassette
AC	acetyl transferase
AD	adenyl transferase
ANS	6-anilino-1-naphthalenesulfonic acid
CAT	chloramphenicol acetyltransferase
CHO	Chinese hamster ovary
CNS	central nervous system
CSF	cerebrospinal fluid
DAB	diaminobutyric acid
DFMO	DL-α-difluoromethylornithine
DHF	dihydrofolic acid
DHFR	dihydrofolate reductase
DHPS	dihydropteroate synthetase
DNA	deoxyribonucleic acid
DON	diazo-oxo-norleucine
dsDNA	double-stranded deoxyribonucleic acid
dTMP	deoxythymidine monophosphate
EDTA	ethylenediaminetetracetic acid
EF	elongation factor
EP	enzyme–product complex
ES	enzyme–substrate complex
FP	ferriprotoporphyrin IX
G6PD	glucose 6-phosphate dehydrogenase
K_i	binding affinity of an inhibitor
Km	Michaelis constant
LTA	lipoteichoic acid
MDR	multidrug resistance
MIC	minimum inhibitory concentration
M_r	molecular mass
mRNA	messenger RNA
MRSA	methicillin-resistant *Staphylococcus aureus*

NADPH	nicotinamide adenine dinucleotide phosphate
NADPH$^+$	oxidized form of NADPH
NAG	N-acetylglucosamine
NAM	N-acetylmuramic acid
NPN	1-N-phenylnaphthylamine
ODC	ornithine decarboxylase
pABA	p-aminobenzoic acid
pAS	p-aminosalicylic acid
PBP	penicillin binding protein
PEP	phosphoenolpyruvate
poly (U)	poly-uridine
PT	phosphotransferase
rbs	ribosome binding site
RNA	ribonucleic acid
rRNA	ribosomal RNA
SDS	sodium dodecylsulfate (sodium lauryl sulfate)
SDS–PAGE	sodium dodecylsulfate–polyacrylamide gel electrophoresis
THF	tetrahydrofolic acid
tRNA	transfer RNA
UDP	uridine diphosphate
Und-P	undecaprenyl phosphate
Und-P-P	undecaprenyl pyrophosphate
UV	ultraviolet
V_{max}	maximum velocity

Preface

The chemical and intellectual background to the era of effective antibacterial therapy was established at around the turn of the 20th century. The early chemotherapeutic drugs were toxic and had to be used with great care. The discovery of the sulfa drugs was a revelation and a stimulus in the 1930s and 1940s but then the naturally occuring antibiotics were discovered and hailed as wonder drugs. Since then many classes of antibiotics have been exploited and significant new families of chemotherapeutic synthetic drugs, especially the quinolones, have been developed. The mechanisms of action of most of these compounds have been worked out in some detail.

The progressive development of resistance to these drugs has been noticed, often soon after the first use of particular antibiotics. The rate of increase in resistance varies from compound to compound, and in some cases strategies have been developed to combat this problem. This has included production of modified forms of the original drug. However the progress of resistance has, at best, been slowed down.

There is now considerable concern that the stock of natural antibiotics that remain to be discovered is running out. These circumstances raise the prospect that infectious diseases may reassert themselves as major causes of sickness and death even in affluent countries.

In such a climate it seems critically important that students understand how antibiotics produce their lifesaving effects, and how the microorganisms may defeat them.

R.A.D. Williams

Chapter 1

Chemotherapy of microbial infections

1.1 Introduction: infections and the history of chemotherapy

1.1.1 Microbial infections

The tissues, body fluids and surfaces of humans and animals provide suitable environments in which micro-organisms may grow. These include the prokaryotic bacteria (formerly known as eubacteria), but not the prokaryotic archaea (formerly known as archaebacteria). Pathogenicity is the ability of a micro-organism to cause disease in its host. In addition to the bacteria, pathogens also include a wide range of eukaryotes, such as fungi, yeasts (Chapter 6) and protozoa (Chapter 7) as well as noncellular 'organisms', the viruses. These last are infectious particles composed of either DNA or RNA, enclosed in protective coats called capsids. The viruses need the materials, enzymes and the synthetic apparatus of a cellular host in order to complete their life cycles.

Many micro-organisms are harmless commensals, including much of the normal microbial flora of the genitalia, the gastrointestinal tract and the skin. The density of micro-organisms varies, feces containing the highest density, at 10^{11} cells g^{-1} (this includes 400 identified species). The total number of micro-organisms carried by man probably exceeds the number of cells in the human body. Some of the normal microbiota of man may be beneficial, if only because they occupy niches and so deny these to more dangerous micro-organisms. However, some members of the normal flora may cause damage to the host over a relatively long period of time, as in the case of dental caries and periodontal disease. Some individuals may carry micro-organisms of proven pathogenic potential for many years without showing any symptoms of disease. The AIDS crisis has dramatically demonstrated

1

the role of the immune system in controlling infections by micro-organisms that are not usually pathogenic in the healthy individual, but which become pathogenic when the immune system is suppressed.

There are a number of features that protect humans and animals from infection of the tissues and body fluids. The immune system normally responds to invasion of the tissues, and copes adequately with many infections. The enzyme lysozyme occurs in tissue fluids and oronasal secretions, where it hydrolyzes the peptidoglycan (Section 3.2) of the bacterial cell wall, and so helps to control the colonization of the body by lysozyme-sensitive bacteria. Sensitivity to this enzyme is governed by whether or not it has access to the peptidoglycan. Other features that minimize the microbial colonization of human tissues include the main-tenance of rather low concentrations of the essential nutrients for micro-organisms in human tissue fluids. For example, iron is avidly bound to proteins such as lactoferrin and transferrins [1], and is not available to most bacteria in this form. Some pathogens have developed the ability to procure iron from the iron-binding complexes of the body. Thus, neisseria have receptor proteins for the iron transporters, and hemolytic bacteria such as the hemophili acquire iron from hemoglobin, and also from heme complexed with haptoglobin and hemopexin [2].

1.1.2 Virulence

Virulence is defined differently in different systems [3], but generally comprises infectivity (the capacity to colonize the host) and also the severity of the disease caused in the host. Highly virulent strains cause high morbidity and perhaps high mortality. Virulence depends on the interaction between the virulence determinants of the parasite and the defensive properties of the host. As the host develops its anti-bodies against the infectious agent by clonal selection, the parasite may simultaneously evolve in order to avoid the immune response. Trypano-somes, *Neisseria meningitidis*, *Salmonella typhimurium* and *Streptococcus pyogenes* are all adept at evolving within the host in this way [4].

1.1.3 The prevalence of infectious diseases and the importance of infections in the past, at present and in the future

The infectious diseases have been major causes of human death through-out history. Major epidemics have occurred periodically and have sometimes caused substantial reductions in population. For example, the great plague of 1664 claimed 15–21% of the population of London. Family sizes were large in Western Europe within living memory, and it was not uncommon for several children in a family to die of infection in one year. The breakdown of societies as a result of wars, famines, economic failure and political upheaval has often been associated with epidemics of infectious diseases, and this continues to be the case.

Improvements to the supply of drinking water and the disposal of sewage have probably made the greatest contribution to the reduction of infections. The era of freedom from the fear of death caused by infection is both recent and brief, as is the ability to prevent and cure infections, at least in affluent societies. In a short period of time the belief that death due to infection is a thing of the past has become widespread. Unfortunately, this happy era may not last. About half of the major companies in Japan and the USA reduced or stopped their antibacterial drug development programs in the late 1980s, and as a result there are few new drugs that can be used in practice in the forseeable future [5]. The resurgence of infectious diseases due to the development of resistance to the available drugs is a real threat to the health of even the most affluent societies.

1.1.4 Development of chemotherapeutic drugs against infections

The earliest known chemotherapeutic agents were of plant origin. The Greeks employed extract of male fern to treat worm infestations, and extracts of cinchona bark were used by South American Indians to treat malaria. The first materials that were not of plant origin which were used chemotherapeutically included mercury, for the treatment of syphilis. Until the beginning of the twentieth century such preparations were the only chemotherapeutic agents available.

At the turn of the century, Ehrlich developed the concept of cell-surface receptors as an extension of the differential staining of tissues by dyes used in histology. Receptor sites on cells recognize particular chemicals, which bind to them, sometimes with great avidity. Ehrlich argued that various cells might have distinct receptors to which different chemicals would bind selectively. He reasoned that toxic chemicals which bound to receptors on pathogens, but not to the host's cells, would be ideal drugs for the treatment of infections. Ehrlich's arsphenamine (Salvarsan, an arsenical drug) was used to treat syphilis until it was superseded by penicillin in the 1940s. Research on dyestuffs as specific chemotherapeutic agents led to the development of the 'sulfa drugs' or sulfonamides (Section 2.3).

In 1928, Fleming observed that colonies of the fungus *Penicillium* inhibited the growth of certain bacteria on plates. Ten years later, Chain and Florey proved that purified penicillin inhibited bacteria infecting mice as well as in cultures *in vitro*. In 1941 the effectiveness of penicillin was demonstrated in man, and the era of antibiotic chemotherapy had begun. The search for other antibiotics was undertaken in many centers, and a great many substances with very diverse structures and modes of action have been discovered. Some of these have been utilized in the chemotherapy of infections in humans and animals.

The introduction of drugs that combat bacterial infections has been a development without parallel in medicine, but it did not occur overnight. The sulfonamides were introduced in 1936, penicillin in 1941 (after its accidental discovery), followed by streptomycin in 1945 (after the first purposeful screening), cephalosporin in 1955, rifampicin in 1959, nalidixic acid in 1962 and cephamycins in 1972.

1.1.5 *The problem of resistance*

The number of antimicrobial agents available was recently so large that it seemed likely that at least one of them would be suitable for all infections. This suitability has been reduced and may be lost altogether by the development of resistance. The wide range of available choices may cause confusion and result in the prescription of drugs that are inappropriate. The tendency to overprescribe is due to the use of antibiotics to satisfy the expectation of patients that they should receive an antibiotic, the use of antimicrobial drugs without sound evidence that a bacterial infection actually exists, and the use of such preparations without proof that a bacterial infection is sensitive to the drug being prescribed, as well as their prescription for viral infections in order to 'cover against secondary bacterial infections'.

The major problem with overprescription, apart from the waste of money involved, is that the more the microbial population is exposed to any inhibitory drug, the more likely it is that resistance will develop. Therefore, unnecessary prescribing should be reduced so that established drugs may retain their effectiveness. A more complex issue is the nonmedical use of antibiotics. Huge amounts of drugs have been included in the diet of farm animals in order to improve the conversion rate (of food into animal products, i.e. eggs, milk or meat). The improved food production may benefit mankind generally, and it has certainly benefited the agriculture industry, but if there is a major resurgence of infectious disease it may prove to have been a profound mistake.

The cost of drugs is a major problem in counteracting resistance. Tuberculosis is ideally treated with three or four drugs over a 2-month period. This is not financially possible in many countries, where treatment may take place over a shorter period, involving fewer drugs and without the support of microbiology laboratory facilities to determine sensitivity [5].

Related to these concerns about the continued effectiveness of drugs is the worry that many of the available antimicrobial drugs that exist in nature may already have been discovered. The more recently discovered drugs may also be more expensive, and yet not necessarily more effective, than were the earlier drugs before resistance to them developed. It has been suggested that the pharmaceutical industry and governments should make more funds available for research in this area, and that they should publicize drugs under development, but not yet under license, that might be used in the event of an emergency [6].

1.2 Some definitions

Chemotherapy is the study and practice of the use of drugs that inhibit or kill an invading species while causing minimum damage to the host. Since this definition, by Ehrlich, the term chemotherapy has been applied to the treatment of infection by unicellular and multicellular organisms, and also to the drug treatment of cancer and viral infections. Only those agents that are effective against micro-organisms will be considered here, but the principle of selective toxicity is the basis of all these forms of chemotherapy. This principle aims to exploit differences between the biochemistry of host tissues and that of the infectious agents, or differences between normal cells and cancer cells. Ideally, the host's biochemical processes of metabolism and biosynthesis should be unaffected by chemotherapeutic drugs. In practice, the ratio of therapeutic to toxic effects, representing the degree of discrimination, should be as high as possible.

Sterilization is the destruction of all microbial life by chemical or physical means. Disinfectants include toxic chemicals such as phenols, formaldehyde and sodium hypochlorite, which may be used in places with dense microbial populations. Ethylene oxide vapor may be used to sterilize apparatus or rooms, and β-propiolactone is a chemical sterilizing agent for solutions, which breaks down by hydrolysis to form a nontoxic acid. Dry heat (150–160°C for 1 h) and moist heat (121°C for 15 min) are commonly used to sterilize equipment, and high-intensity γ-irradiation is used to sterilize disposable plasticware and needles.

Sanitizing agents, antiseptics or germicides (these terms are used interchangeably) may not sterilize completely, but they do reduce the microbial population to acceptable levels. They are milder than the disinfectants, and so are suitable for use on living tissue. They include dilute solutions of chlorinated phenols, some ionic detergents and the flavines used in wound dressings. Antimicrobial agents are those chemotherapeutic compounds, whether synthetic or natural, that are toxic to micro-organisms. Antibiotics are chemicals produced by micro-organisms (some bacteria, fungi and actinomycetes) that are highly toxic to other micro-organisms at low concentrations. Some antibiotics can be synthesized, but most are made by culturing the producing micro-organisms on a large scale under closely controlled conditions. Antibiotics represent the majority of the antimicrobial drugs, and are

Table 1.1: The use of major antibiotics in a large teaching hospital

Antibiotic	Amount used per year (kg)
Penicillins	121.0
Erythromycin	10.1
Cephalosporins	8.1
Aminoglycosides	2.1
Fusidate	1.4
Tetracyclines	1.3
Chloramphenicol	0.6

used in very large quantities. *Table 1.1* indicates the quantities of major antibiotics described in this book that were supplied by the pharmacy of The Royal London Hospital during a single year.

Some antibacterial agents are bactericidal, (i.e. they kill the bacteria), but often it is not necessary to kill all the infective micro-organisms during the treatment of disease. Bacteriostatic compounds prevent further growth of the bacteria, allowing the host's normal defense mechanisms to clear up the infection. The two types of action may not be clearly distinguishable because the test of whether bacteria are alive is their successful growth. It is important to remember that many chemotherapeutic agents do not eradicate all the infective organisms, but they allow the body defenses to do this, and thus normal immuno-logical and phagocytic mechanisms must be working efficiently if such drugs are to be effective. It has become more important to be aware of this difference, now, at a time when increasing numbers of patients are immunocompromised. It is therefore important that patients complete their course of drugs in order to avoid the persistence of bacteria that have an enhanced likelihood of resistance.

1.3 Targets of drug action

The sites of action against which the chemotherapeutic antimicrobial drugs have been directed are diverse (*Table 1.2*), but the most widely useful drugs have been directed against the synthesis of peptidoglycan of the bacterial cell walls (Chapter 3) and bacterial protein biosynthesis (Chapter 5). Other significant targets include intermediary metabolism involving the folate coenzymes (Chapters 2 and 7), the biosynthesis of DNA and RNA (Chapter 4) and cell membranes (Chapter 6).

1.4 Sensitivity of bacteria and acquired and infectious resistance

Bacteria differ in their sensitivities to any particular antimicrobial drug, and the determination of sensitivity is important in the reduction

Table 1.2: Targets for the activity of antimicrobial agents

Cell wall: the synthesis of peptidoglycan
Cytosolic phase
Membrane phase
Extra-membranous phase
Cell membrane function
Intermediary metabolism
Anti-folate compounds
Protein synthesis
Aminoacyl-tRNA synthesis
Ribosome function
Nucleic acid synthesis
Antimetabolites inhibiting early stages
Intercalators
Anti-folate compounds
RNA polymerase
DNA gyrase

of inappropriate prescriptions and in minimizing the development of resistant strains. One way to reduce the incidence of resistant strains is to give two different antibiotics, so that each drug inhibits any mutants that are resistant to the other. This tactic is less useful against the infectious spread of drug resistance factors (plasmids) in a bacterial population, since these factors often confer resistance to several anti-biotics simultaneously.

1.4.1 Intrinsic resistance

Factors that restrict the usefulness of antibiotics include the initial sensitivity of the infecting micro-organisms, and the ways in which this may change after exposure to the antibiotic. Intrinsic resistance may occur because the drug cannot gain access to its site of action in the bacteria. An important example of this is the failure of many penicillins to inhibit Gram-negative bacteria. These micro-organisms have an outer hydrophobic cell wall layer which many penicillins cannot penetrate. The semi-synthetic penicillins with hydrophobic side-chains may pass through this outer wall layer and so be active against a broader spectrum of bacteria than the unmodified penicillins.

Enterococci (formerly known as streptococci) are a major cause of nosocomial (hospital) infections, and they show a wide innate resistance to drugs, including cephalosporins and semi-synthetic penicillins (Section 3.8.2), clindamycin and low concentrations of aminoglycosides (Section 5.4.1 and *Table 1.3*). The glycopeptides, such as vancomycin, have been among the few drugs available over the past 30 years to treat infections with enterococci, particularly by β-lactamase producers, that destroy penicillins, and methicillin-resistant *Staphylococcus aureus*

(MRSA). The recent development of vancomycin resistance among the enterococci is of great concern because of the wide inherent resistance of the enterococci, and the possibility that the resistance may spread to other pathogens. *Pseudomonas aeruginosa* is an important pathogen that is noted for its inherent resistance to many of the clinically useful antibiotics (*Table 1.3*).

1.4.2 Resistant mutants

Bacteria that are exposed to antibiotics to which they are sensitive are either killed, or their growth is suppressed. Spontaneous mutants that have any character which confers resistance to the drug will be selected by the drug, and clones of such resistant mutants will proliferate. For this reason, any unnecessary use of antibiotics that increases the exposure of bacteria to this selective force should be avoided.

The mechanisms of development of resistance are as varied as the mechanisms of action of antibiotics themselves. In general terms, these mechanisms include the development of impermeability to the drug, changes in the binding of drugs to their sites of inhibition, and the induction of enzymes that modify or destroy the drugs themselves.

1.4.3 Infectious drug resistance

The transmission of resistance to antibiotics from one bacterium to another was attributed to 'R factors' which often transmitted resistance to several drugs simultaneously. These factors are plasmids, namely small self-replicating double-stranded DNA circles, which are independent of the chromosome of the host bacterium. Plasmids may occur in few or many copies in a bacterium, and may have very specific or wide host ranges. Bacteria can usually be 'cured' (i.e. exposed to conditions that cause the loss of the plasmids). Although genes other than those for antibiotic resistance may be carried, bacteria usually survive without plasmids, perhaps with a modified phenotype if the plasmids bear genes.

1.5 Sensitivity testing of bacteria

Whenever possible, samples of sputum, blood, feces, urine or pus should be taken before any drugs are given. The laboratory diagnosis of the sensitivity of micro-organisms to antimicrobial agents is important, although therapy with a broad-spectrum drug will usually be started before these results are available. Sensitivity results are usually available within a few days from laboratories which use classical techniques for testing. These include the use of agar plates to determine the size

Table 1.3: Susceptibility of some micro-organisms to selected antibiotics from the Glaxo 'Microbe Base' database

Antibiotic	Organism	No. isolated	No. tested (%)	Sensitive (%)
Ampicillin	*Staphylococcus*	43 156	10 120	15.1
Methicillin	*aureus*		8865	99.6
Flucloxacillin			32 716	97.8
Vancomycin			6983	99.9
Trimethoprim			11 977	83.2
Erythromycin			39 917	92.1
Fusidate			28 341	96.9
Tetracycline			17 918	90.2
Ampicillin	*Escherichia coli*	74 813	51 112	59.1
Ceftazidime			56 246	99.3
Sulfonamide			33 602	66.3
Trimethoprim			66 391	82.1
Tetracycline			12 142	74.1
Nitrofurantoin			63 529	95.3
Ampicillin	Enterococci	10 204	8783	95.8
Ceftazidime			4849	7.7
Vancomycin			466	99.3
Sulfonamide			2332	12.1
Trimethoprim			6431	74.6
Gentamicin			2096	4.1
Fusidate			237	8.0
Erythromycin			1602	58.2
Tetracycline			904	37.4
Nitrofurantoin			5781	97.3
Ampicillin	*Bacteroides fragilis*	1253	200	7.0
Augmentin			446	97.9
Ceftazidime			255	32.9
Erythromycin			691	63.0
Tetracycline			493	40.5
Metronidazole			1243	99.1
Ampicillin	*Hemophilus influenzae*	8703	7634	87.3
Ceftazidime			4662	97.5
Sulfonamide			2828	87.8
Trimethoprim			5496	90.7
Erythromycin			5297	72.5
Tetracycline			6352	96.4
Ampicillin	*Pseudomonas*	9606	338	4.4
Piperacillin	*aeruginosa*		4598	88.8
Ceftazidime			9101	94.0
Sulfonamide			340	29.7
Gentamicin			8466	90.2
Fusidate			50	92.0
Erythromycin			137	33.5
Tetracycline			232	3.4
Nitrofurantoin			611	2.1

of the zones of growth inhibition around paper discs impregnated with drugs, and determination of the inhibition of bacterial growth in broth. Rapid methods such as conductometric testing for ionic changes associated with bacterial growth in the presence and absence of antibiotics may make sensitivity results available more quickly. The minimal inhibitory concentration (MIC) of an antibiotic is defined as the lowest concentration in a dilution series that completely prevents growth of a test strain.

Databases of the current susceptibilities of micro-organisms are maintained in order to enable clinicians to keep abreast of changes in the drug resistance of the infectious agents that they encounter (*Table 1.3*).

1.6 The effect of bacterial growth rate on susceptibility

The growth rate of bacteria affects their composition, including the constituents of their cell walls. In general, slow-growing cells appear to be inhibited less than fast-growing cells by a range of toxic substances including ethylenediaminetetraacetic acid (EDTA), cationic detergents such as chlorhexidine, and anionic detergents such as sodium dodecylsulfate (SDS). The susceptibility of *P. aeruginosa* to polymyxin increases as growth becomes more rapid, but it is not possible to relate this change to any one cell-wall component [7]. In the case of *Escherichia coli* the effect of growth rate in continuous culture depends on the limiting nutrient. The susceptibility of bacteria to β-lactams is very dependent on growth rate. Slow growth of cells precludes lysis, which is the cause of cell death.

1.7 The post-antibiotic effect

The growth of bacteria may be repressed for an extended period of time after even a brief exposure to an antimicrobial drug. This period of recovery between the removal or inactivation of the antibiotic and the resumption of normal growth may range from 20 min to several hours. The phenomenon is called the post-antibiotic effect, and it is of relevance to dose regimes. Drug–micro-organism interactions that show the post-antibiotic effect may need more frequent doses of drug than those that do not show this effect. The post-antibiotic effect is produced by β-lactams, aminoglycosides, fluoroquinolones and macrolides. Even though intermittent doses of penicillin result in ineffective serum concentrations over long periods, infections are often resolved. The post-antibiotic effect ensures that the bacteria fail to grow while the plasma penicillin concentration is low. During the phase of suppressed growth the bacteria may be markedly more susceptible to leukocytes.

Table 1.4: Adverse effects of some antimicrobial agents[a]

Tissue	Drug	Effects and comments
Liver	Isoniazid	Often in rapid acetylators
	Rifampicin	Usually mild
	Tetracyclines	Huge doses
	Erythromycin estolate	
	Nitrofurantoin	
	β-Lactams	Brief rise in serum aminotransferases
Kidney	Aminoglycosides	
	Cephalosporins	
	Sulfonamides	
	Methicillin	
	Amphotericin B	
	Polymyxins	
Intestine	Penicillins	Diarrhea – especially ampicillin
	Tetracyclines	Diarrhea
	Clindamycin	Pseudomembranous colitis
Nervous system	Aminoglycosides	8th nerve and neuromuscular block
	Metronidazole	Peripheral neuropathy
	Nitrofurantoin	Peripheral neuropathy
	β-Lactams	Convulsions – large intrathecal dose
Bone marrow	Amphotericin B	Suppression
	Chloramphenicol	
	Sulfonamides	
Hemolytic anemia	Sulfonamides,	
	Nitrofurantoin	
	Quinolones	
	β-Lactams	Rare
Thrombocytopenia	Sulfonamides	Rare
	Cephalosporins	Rare
	Rifampicin	On intermittent therapy
Neutropenia	β-Lactams	Rare
	Sulfonamides	Rare
	Chloramphenicol	

[a]Data reproduced from Reid *et al.* (1992). *Lecture Notes on Clinical Pharmacology*, with permission from Blackwell Science Ltd.

1.8 Side-effects of chemotherapeutic drugs

Antimicrobial drugs are intended to kill or incapacitate an infective micro-organism, and any other effect on the host that may be undesirable or toxic is a side-effect. The effects of antimicrobial drugs that are unconnected with the mechanism of action against the primary target are very diverse, and may severely restrict the use of a drug (*Table 1.4*).

Toxicity to the host includes hypersensitivity (allergic) reactions in persons who (possibly unknown to themselves) have had previous contact with the drug. This may involve the drug, or unidentified constituents of the drug preparation, or their reaction products in the body, acting as antigens. Antibodies are produced in the body more profusely when further exposure to the antigen occurs. The response can vary from mild rashes to a fatal anaphylactoid reaction. A characteristic of hypersensitivity reactions is that they are independent of the dose of drug administered. It is prudent to err on the side of caution when hypersensitivity is suspected, and to immediately withdraw the drug and ensure that both the patient and those who are caring for him or her are aware that the drug or any close congeners must not be used in future.

Forms of toxicity other than hypersensitivity reactions are often dose-dependent, and the likelihood of toxic effects can be predicted, so it should be relatively simple to avoid toxic concentrations. Certain forms of drug toxicity are unpredictable and idiosyncratic. Fortunately such reactions are rare, but they can be very serious. Toxic concentrations of drugs are particularly likely to occur when drug elimination is impaired, as in liver or kidney failure.

1.9 Pharmacological considerations

The pharmacokinetics of antimicrobial drugs consist of the way in which the body deals with the drug and the actions of the body on the drug. Administration may take place by different routes and with a frequency that may vary considerably (e.g. from once daily [8] to continuous infusion [9]).

1.9.1 Routes of administration and drug absorption

Oral administration is effective in gastrointestinal infections, in which case absorption is neither necessary nor desirable. For systemic action a drug must be lipid-soluble in order to be absorbed from the gastrointestinal tract. Substances that are not lipid-soluble may be partly absorbed from the gut, but may be more effectively administered by injection.

Drugs that are unstable in the presence of stomach acid (penicillin G, erythromycin) are poorly absorbed if given orally. Most drug absorption occurs by passive diffusion in the small intestine. The lipid solubility of a drug is influenced by pH, the unionized form is more lipid-soluble, and for optimal absorption it is best for the drug not to be ionized. Compounds with acidic groups usually have pK_a values below pH 7, and are less ionized at low pH values. Substances derived from bases usually have pK_a values above pH 7, and are less ionized at high pH values.

Parenteral injection is used for drugs that cannot be given by another route. Intravenous injection provides a rapid onset of action because the absorption stage of drug intake is circumvented. Absorption of a drug from an injection depends on its lipid solubility and on the blood supply to the injection site. Intrathecal injections are occasionally administered in life-threatening situations, or when a rapid onset of action in the central nervous system is required, or if multiple painful injections would otherwise be necessary.

Drugs may be applied topically to the skin or to the eye when a purely local action is required. High local concentrations can be achieved in this way, but local adverse effects and absorption from the site may occur. A drug that acts on the respiratory system may be given by inhalation as an aerosol or in the form of fine particles, so that it penetrates the small bronchi. Inhalation is inappropriate if the bronchi are blocked, as the drug may not reach the site of infection. Careful attention must be given to the dose administered by this route, as absorption may result in toxic effects.

Formulation affects the ease with which a drug is absorbed. Microfining to a small particle size facilitates absorption.

1.9.2 *Drug distribution*

The lipid solubility and degree of ionization of a drug affect the distribution of the drug across the membranes in the body. The 'blood–brain barrier' is a lipid interface that is readily crossed by lipid-soluble drugs, and the permeability of the blood–brain barrier is increased when the meninges are inflamed, partly due to increased blood flow.

Effective concentrations of a lipid-insoluble drug may occur in tissues if high plasma concentrations are achieved by intravenous injection, but toxic effects may also become more pronounced. A good blood supply helps a drug to reach its target site at an effective concentration. In order to clear an infection in a tissue with a poor blood supply, it may be justifiable to use higher doses of the drug than would be employed for a highly vascular tissue.

The concepts of tissue penetration by antibiotics have been reviewed previously [10, 11]. Drugs may be transported in the blood both in solution in the plasma and bound to the blood constituents, most commonly to the plasma proteins. A compound dissolved in the plasma is available for transport to the site of action, but drugs that are combined with other blood constituents may not be totally available. Plasma-protein binding, in equilibrium with the drug in solution, provides a reserve supply of drug which extends the time over which a given antibiotic concentration is maintained. The concentration of an antimicrobial drug in plasma or serum is considered to provide a good indication of whether administration is adequate, and 'effective' ranges of plasma concentration have been established for many antimicrobial drugs.

When two drugs that compete for the same plasma–protein binding site are given simultaneously, the drug with the higher affinity will bind preferentially, and more of the other drug will be available in a free form. Abnormally high or toxic concentrations of the unbound drug may then occur. Such an interaction is exemplified by nalidixic acid, which displaces warfarin from plasma proteins, resulting in prolongation of clotting time and hemorrhage, and sulfonamides, which displace bilirubin from plasma proteins in newborn infants, so causing kernicterus.

1.9.3 Metabolism

Metabolic conversion may increase or decrease the activity of a drug. The major types of reaction involved are oxidation, reduction, hydroxylation and conjugation, mainly by liver microsomal enzymes (microsomes are pinched-off vesicles of endoplasmic reticulum containing cytoplasm). Metabolism usually improves drug excretion because the products are usually both more water-soluble and more readily ionized than the parent drug. Metabolic changes in activity are exemplified by the hydrolysis of erythromycin estolate to give free active erythromycin, and the glucuronidation of chloramphenicol to give an inactive product. Altered liver function can therefore have major consequences for the therapy of infections. Metabolism is slowed in liver failure, or accelerated if the microsomal enzymes have been induced by drugs such as barbiturates, meprobamate, phenylbutazone or ethanol.

If liver function is poor, the concentration of plasma proteins that bind drugs is often decreased, which may result in a higher concentration of free drug.

1.9.4 Excretion

Drugs are usually excreted in the urine or by secretion into the bile. Other routes of excretion, such as via the lungs and the skin, do not play a significant role in the excretion of antimicrobial agents. Only compounds in free solution in the plasma are available for excretion via the kidneys. Drugs that are metabolized to more water-soluble and more ionized products than the parent compound are more rapidly excreted. The major mechanisms of excretion are filtration at the glomerulus and active secretion into the lumen of the kidney tubules. Measurement of creatinine clearance capacity provides a good indication of the functional state of the glomerular filtration process, which tends to reduce with age.

Penicillins are excreted by active secretion into the lumen of the kidney tubules, which can be disadvantageous as it is necessary to administer the drug frequently in order to maintain therapeutic concentrations. Penicillin excretion can be reduced by probenecid, which slows down the transport of penicillin across the renal tubules.

Drugs that are actively secreted by the liver into the bile frequently have a molecular weight in excess of 300, and many substances that are glucuronidated are excreted via this route. Unless reabsorption of the drug occurs in the small and large intestines, it is excreted in the feces. However, if some reabsorption of the drug occurs at these sites, the drug may exert its biological effects again, before it is secreted into the bile once more. This enterohepatic circulation may involve a drug molecule making several secretion and reabsorption cycles before it is finally lost from the body.

1.9.5 Effects of age, liver function and kidney function

The quantity of drug administered is usually based on the dose that has been found to be empirically effective in 'the normal 70-kg man'. Such doses may be inappropriate in many cases for a variety of reasons, including extremes of age and body weight, and changes in liver and kidney function.

Age. In the very young, many biological functions are poorly developed. Acid secretion in the stomach, microsomal enzyme activity in the liver, and glomerular filtration and tubular secretion in the kidney are all poorly developed in the premature baby and in the neonate. In addition, the blood–brain barrier does not appear to be fully functional very early in life. Therefore, in the very young the dose of drugs should be reduced. In older children and young adults, drug dosage is usually determined on the basis of the ratio of body weight to 70 kg. The elderly may have a reduced ability to secrete stomach acid, as well as some reduction in liver and kidney function. This may result in an increased ability to absorb acid-labile drugs, and a decreased ability to metabolize and excrete them, all contributing to higher than normal plasma concentrations which may occur in the elderly if the drug dosage is not adjusted.

Liver function. The inability to metabolize a drug, or to secrete it into the bile, may cause it to accumulate in the body. As the toxicity of some drugs is related to their concentration in the plasma, toxicity may result from decreased liver function. Changes in drug response may also be caused by the decrease in plasma proteins which occurs during liver failure, when abnormally high concentrations of the free drug may occur. Therefore it may be necessary to reduce the amount of drug given, and to increase the frequency of administration. Monitoring the plasma concentration of drugs is recommended under such circumstances.

Kidney function. The urinary excretion of drugs or their metabolites can be significantly delayed during kidney failure, and if the frequency of administration is not reduced, then toxic effects may rapidly develop. The quantity of drug administered during kidney failure should not be reduced as this could result in inadequate plasma concentrations.

The duration of action of a drug in the body depends on the physico-chemical characteristics of the drug, the formulation, the route of administration, and not least the way in which the drug is handled in the body. The half-life or plasma half-time is a measure of the time taken for the plasma concentration of a drug to be reduced by 50%. The half-time must not be assumed to be the same as the duration of action of the drug, as drugs which become irreversibly bound to tissues can be effective long after they are undetectable in plasma.

Measurement of the plasma concentration is the best method available for estimating the amount of drug in the body. Comparison of the plasma concentration of an antimicrobial drug with the MIC (minimum inhibitory concentration) for a pathogenic micro-organism provides a good indication of whether the antimicrobial drug will effectively combat an infection.

1.9.6 Drug interactions

If two or more drugs administered together share a common site of action, the activity of one or both of them may be either increased or decreased. The absorption of other drugs is reduced by opiates and barbiturates that depress intestinal motility. Tetracyclines chelate divalent metal ions such as Ca^{2+} and Fe^{2+}, which tend to retain them in the gut lumen. The presence of food in the intestine delays drug absorption, and alteration of pH affects the ionization and lipid solubility of drugs. During distribution in the body, competition for plasma–protein binding sites by sulfonamides may displace bound tolbutamide and so cause exaggerated hypoglycemic responses.

Drug metabolism may be reduced in liver failure, and by monoamine oxidase inhibitors, leading to exaggerated responses. By contrast, the induction of microsomal enzymes by barbiturates and phenytoin leads to more rapid metabolism of drugs. Changes in urinary pH affect the excretion of drugs. More specifically, probenecid inhibits the active secretion of penicillin in the kidney tubules.

If a drug interaction produces a response that is greater than the sum of the individual effects of the drug, the interaction is synergistic, and if the response is less than that of the compounds alone, then the interaction is antagonistic. An important example of synergism is the interaction of sulfamethoxazole and trimethoprim in the combination known as cotrimoxazole. Sulfamethoxazole and trimethoprim act at successive steps in the intermediary metabolism of folic acid in bacteria. When administered by themselves, high doses of each are

The Healthcare Library of Northern Ireland and
Queen's University Biomedical Library

Customer ID: ****4182

Items that you have checked out

Title:
:Clinical pharmacy and therapeutics / edited
by Cate Whittlesea and Karen Hodson.
ID: 70906904
Due: 29 April 2019

Title:
Antibiotics simplified / Jason C. Gallagher,
Conan MacDougall.
ID: 69235694
Due: 29 April 2019

Title:
Antimicrobial drug action / R.A.D. Williams,
P.A. Lambert, P. Singleton.
ID: 64713695
Due: 29 April 2019

Title:
Churchill's Pocketbook of clinical pharmacy /
edited by Nick Barber, Alan Willson.
ID: 67867731
Due: 29 April 2019

Title: Pharmacotherapy handbook / Barbara G.
Wells ... [et al].
ID: 69698481
Due: 29 April 2019

Total items: 5
29/03/2019 15:49
Checked out: 7

Check the DUE DATE! Fines accumulate at a
rate of 10p per item per day

(Biomed)

needed in order to eliminate bacterial infection. When given together, however, the dose of each drug can be reduced to obtain the same chemotherapeutic effect. Synergism can be of therapeutic significance as the risk of toxicity is reduced by the lower dosages of the two drugs used.

1.10   Diarrhea and super-infection by bacteria after antibiotic therapy

During oral administration of certain broad-spectrum antibiotics (e.g. tetracycline), most or all of some bacterial species in the intestine may be killed. When this occurs, other species in the gastrointestinal tract, which are not normally pathogenic, may multiply in an uncontrolled manner and result in super-infection. This is usually characterized by diarrhea, and can be life-threatening if not treated [12]. Under such circumstances, it is necessary to identify the species responsible for the super-infection, and to treat the patient with a drug to which this species is sensitive. Although the micro-organism(s) responsible may be unidentified (*Table 1.5*), the classic example of super-infection is the pseudomembranous colitis due to *Clostridium difficile* after treatment with clindamycin, ampicillin or cephalosporins or antitumor agents.

Table 1.5: Antibiotic-associated diarrhea and colitis due to *Clostridium difficile* and to unknown causes

Feature	*Clostridium difficile*	Unknown micro-organisms
Drugs commonly implicated	Clindamycin, ampicillin, and cephalosporins	Clindamycin, ampicillin, some cephalosporins, tetracycline
Relationship to dose	None	Dose-related
Response to drug withdrawal	Symptoms may persist	Symptoms usually resolve
Feature	Watery diarrhea, cramps	Watery diarrhea
Complications	High fever, dehydration, toxic colon, ileum, hypoalbuminemia	Rarely serious
Colitis	Cramps, leukocytes in feces	Uncommon
Epidemiology	Epidemic/endemic in hospitals and nursing homes	Sporadic
Discontinue drug	Yes	Yes, or reduce dose
Treatment	Vancomycin or metronidazole	Vancomycin sometimes

References

1. Welch, S.G. (1993) *Transferrin: The Iron Carrier*. CRC Press, Boca Raton.
2. Briat, J.-F. (1992) *J. Gen. Microbiol.*, **138**, 2475–2483.
3. Read, A.F. (1994) *Trends Microbiol.*, **2**, 73–76.
4. Levin, B.R. and Bull, J.J. (1994) *Trends Microbiol.*, **2**, 76–81.
5. Border, P. (1994) *Diseases Fighting Back – The Growing Resistance of TB and other Bacterial Diseases to Treatment*. Parliamentary Office of Science and Technology Report, POST, London, 1–36.
6. Tomasz, A. (1994) *N. Engl. J. Med.*, **330**, 1247–1251.
7. Nix, D.E., Goodwin, S.D., Peloquin, C.A., Rotella, D.L. and Schentag, J.J. (1991) *Antimicrob. Agents Chemother.*, **35**, 1947–1952 and 1953–1959.
8. Brown, M.R.W., Collier, P.J. and Gilbert, P. (1990) *Antimicrob. Agents Chemother.*, **34**, 1623–1628.
9. Gilbert, D.N., Collier, P.J. and Brown, M.R.W. (1990) *Antimicrob. Agents Chemother.*, **34**, 1865–1868.
10. Gilbert, D.N. (1991) *Antimicrob. Agents Chemother.*, **35**, 399–405.
11. Craig, W.A. and Ebert, S.C. (1992) *Antimicrob. Agents Chemother.*, **36**, 2577–2583.
12. Bartlett, J.G. (1992) *Clin. Infect. Dis.*, **15**, 573–579.

Chapter 2

The antimetabolites: folate metabolism as a target for antimicrobial drugs

2.1 Introduction to inhibition of cell metabolism

The drugs described in this chapter are effective against the enzyme-catalyzed reactions of intermediary metabolism in micro-organisms, but they ultimately affect the synthesis of macromolecules. They may also be used against protozoal infections (Chapter 7).

The substrate in an enzyme-catalyzed reaction binds to its enzyme to form a transient enzyme–substrate complex (ES). This complex proceeds, perhaps through several stages, to give the enzyme–product complex (EP), which then dissociates. The velocity of the enzyme-catalyzed reaction increases as the substrate concentration rises, until all the enzyme molecules are saturated with substrate. It then reaches a plateau at the maximum velocity (V_{max}), which is dependent on the concentration of enzyme molecules. The affinity of an enzyme for its substrate is measured by the Michaelis constant (K_m), which is the concentration of substrate that gives half the maximum (saturation) velocity (V_{max}) for any amount of enzyme. A low K_m value means the enzyme is effective at low substrate concentration (i.e. it has a high affinity for the substrate). The K_m can be determined by studying the effect of substrate concentration on the reaction velocity.

Drugs that show structural similarity to the normal substrate of an enzyme may bind to the active site of the enzyme in place of the substrate, but do not react to give a product. Such compounds are competitive inhibitors, and the degree of inhibition depends on the concentration of the substrate, the concentration of the inhibitor and the affinities of each for binding to the enzyme. The binding affinity of an inhibitor is defined by K_i which can be determined indirectly by determining the K_m in the absence of inhibitor and the apparent K_m values in the presence of various concentrations of the inhibitor.

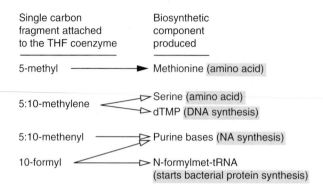

Figure 2.1: The importance of folates in metabolism and as targets for chemotherapy.

2.2 Folate metabolism

Folate is a coenzyme essential for cell growth, but most bacteria cannot transport folate, and must manufacture it *de novo*. By contrast, humans cannot make folate and must ingest it as a vitamin. Folate biosynthesis is thus an ideal target for antimicrobial drugs. The metabolic interactions of folates that are important in chemotherapy are indicated in *Figure 2.1*. There are two energy-dependent folate transport mechanisms in mammals [1], but very few bacteria (some species of *Lactobacillus* and *Streptococcus*) require folic acid as a vitamin.

2.3 The discovery of sulfonamides

The studies of Ehrlich on the selective binding of dyes to cells led to the discovery of the sulfonamide drugs. Prontosil rubrum, a diazo dye, prevented the death of mice infected with hemolytic strepto-cocci. The dye was introduced under the name of Streptozon in 1935, but was only effective *in vivo*. However, a reduction product of the dye, namely sulfanilamide, was effective as a bacterial inhibitor both *in vitro* and *in vivo*. The reductive cleavage to sulfanilamide occurs in the liver (*Figure 2.2*), and it was concluded that sulfanilamide was responsible for the antibacterial action of prontosil, and that any hypo-thetical affinity of the dye for cells was irrelevant to its antibacterial action.

Prontosil and sulfanilamide were used successfully against infections caused by hemolytic streptococci, and the number of deaths from puerperal sepsis (about 1000 per annum in 1935 in England and Wales) was reduced dramatically. The sulfonamide series of drugs are synthetic derivatives of sulfanilamide (e.g. sulfamethoxazole and sulfafurazole) and are not therefore antibiotics. The amino group is

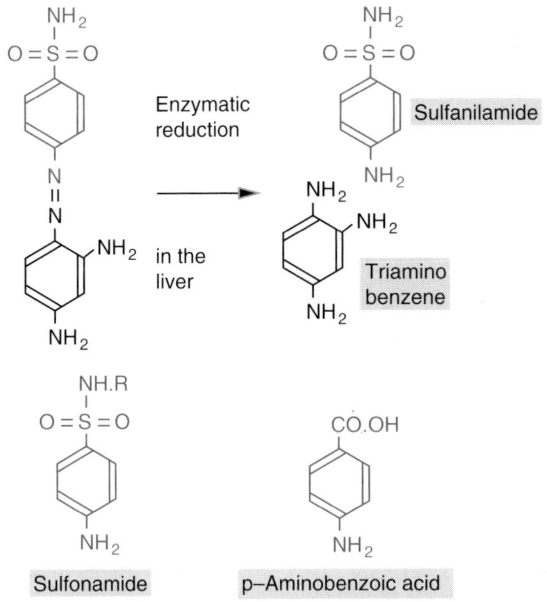

Figure 2.2: Metabolic cleavage of prontosil rubrum in the liver, and the similarity between p-aminobenzoic acid and the sulfonamides.

usually unsubstituted, except in the case of succinyl and phthalyl sulfonamides, which have a long duration of action and are not well absorbed from the gut. Active sulfonamide is slowly released after hydrolytic removal of the succinyl or phthalyl residue in the tissues. Many of the successful sulfonamide derivatives have a heterocyclic substituent. Sulfapyridine (M&B 693) was the first drug against pneumococcal pneumonia.

2.4 The mechanism of action of sulfonamides

Sulfonamides inhibit growing micro-organisms and extracts of bacteria or yeast compete with this antibacterial action. In 1940, Woods suggested that a metabolite or growth factor was similar in structure to sulfanilamide, and synthetic p-aminobenzoic acid (pABA) was highly effective in this role. The coenzyme folate was the essential cell component, the synthesis of which was inhibited by sulfanilamide [2].

Bacteriostasis is preceded by a time lag, the duration of which depends on the quantity of folate coenzymes stored. The killing effect of sulfonamide, 'thymine-less death', has been specifically attributed to cessation of DNA synthesis while protein synthesis proceeds, but the mechanism responsible is unknown.

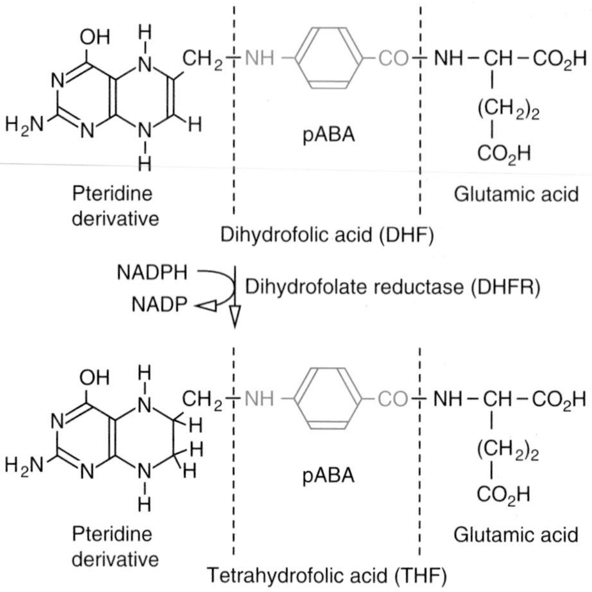

Figure 2.3: The structures of dihydrofolate and tetrahydrofolate, and the role of dihydrofolate reductase.

In sulfonamide-resistant mutants of pneumococcus, the enzyme that catalyzes the reaction of pteridine pyrophosphate with *p*-aminobenzoate, dihydropteroate synthetase, has altered kinetic constants [3]. A sensitive enzyme wild-type had a K_m for pABA of 0.1 μM and a K_i for sulfonamide of 0.5 μM. By contrast, the mutant enzyme from a resistant strain had a K_m of 0.06 μM and a K_i of 65 μM. The increase in the K_i/K_m ratio allows the resistant mutant to discriminate against sulfonamide inhibition at therapeutic concentrations.

Sulfonamides inhibit the incorporation of pABA into a precursor of dihydrofolic acid (DHF) that should then be reduced by the enzyme dihydrofolate reductase (DHFR) to tetrahydrofolic acid (THF) (see *Figure 2.3*). This derivative of folic acid is an important coenzyme involved in the transfer of small residues containing a single carbon atom (e.g. methyl, formyl) in intermediary metabolism. The synthesis of the amino acid methionine and of nucleic acid bases, including thymine, involves the use of THF derivatives (*Figure 2.1*). DHF is synthesized in two stages, the first of which is the combination of the pteridine derivative with pABA, catalyzed by dihydropteroic acid synthetase, the reaction that is inhibited by sulfonamides. This is followed by condensation of glutamic acid with dihydropteroic acid. Because pABA and sulfonamides compete for the same enzyme, the

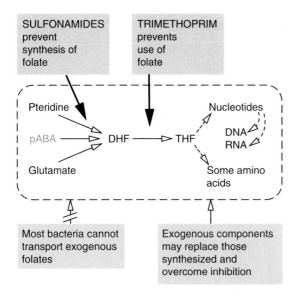

Figure 2.4: The metabolism of folate in bacteria and the targets for chemotherapy by drugs.

inhibitory effects of sulfonamides can be reversed if exogenous pABA is available in sufficiently high concentrations. Bacteria (with few exceptions) do not utilize exogenous folates, which they cannot absorb (*Figure 2.4*). They therefore depend on the synthesis of dihydrofolate within their cells, and if this process is inhibited the formation of macromolecules is impaired. Other drugs (e.g. trimethoprim) that inhibit the conversion of dihydrofolate to tetrahydrofolate also limit the supply of these biosynthetic components. Therefore the time lag between administration of sulfonamides and the cessation of bacterial growth corresponds to the time taken to use up the stocks of these biosynthetic components, and of folic acid already present in the cell. Growth proceeds for about four cell generations after sulfonamide treatment, by which time the original stock of folic acid is diluted amongst the progeny of the treated bacteria.

As well as being competitive inhibitors of this enzyme, sulfonamides may also act as substrates of dihydropteroate synthetase [4, 5], and [^{35}S]sulfanilamide causes the accumulation of labeled analogs of folic acid. If such analogs of dihydropteroic acid are provided exogenously as potential drugs, they are therapeutically ineffective, presumably because they need a folate transporter to enter bacteria, but if they are synthesized inside bacteria they may be expected to have deleterious effects.

2.5   Sensitive bacteria and resistance to sulfonamide

Bacteria that are sensitive to sulfonamides include some strains of streptococci, including pneumococci, some actinomycetes, *Corynebacterium
diphtheriae* and *Hemophilus influenzae*. Because the sulfonamides
inhibit the synthesis of nucleotides and amino acids, the drugs are not
effective if these compounds are available from exogenous sources
(*Figure 2.4*) (e.g. when there is extensive tissue damage, as in burns and
purulent exudates). Most tissue fluids, including blood, have relatively
low concentrations of amino acids and do not contain nucleotide bases
in a form that can be absorbed by bacteria.

Sulfonamides are now of very limited use for the treatment of
bacterial infections (*Table 1.3*). They are sometimes used in combination with another drug, but they have largely been replaced by natural
antibiotics for treatment of streptococcal and pneumococcal infections.
A high proportion of the dose is excreted unchanged in the urine, which
is appropriate for the treatment of urinary tract infections.

The role of sulfonamides in the competitive inhibition of folic acid
synthesis (*Figure 2.4*) implies that resistance may be due to acquired
impermeability of the bacterial membrane to the drugs, or to increased
production of the substrate (pABA) to overcome the inhibition, or to
a change in the properties of dihydropteroic acid synthetase so that the
enzyme binds to the inhibitor less readily. Both of the latter two types
of resistance have been demonstrated. The increased intracellular
concentration of pABA in resistant bacteria may be due to a loss of
the normal end-product inhibition of pABA synthesis. The altered
properties of dihydropteroic acid synthetase have been demonstrated
in a resistant mutant in which the enzyme has a greater affinity for the
substrate and a lower affinity for sulfonamides than the sensitive
enzyme.

Sulfonamides may also act as substrates for dihydropteroic acid
synthetase and produce folic acid analogs that may exert their own
inhibitory effects, but the prevalence and significance of these derivatives is not clear. As with many other antimicrobial agents, it is too
simplistic to assume that there is only one mode of action, and that this
is the same in all bacteria.

2.6   Pharmacology of sulfonamides

2.6.1   Administration

The duration of action after absorption from the gut is determined by
the speed of metabolism and excretion. Sulfafurazole is soluble and
readily excreted, reaching high concentrations in the urine, and is therefore used to treat urinary tract infections. Sulfadiazine and sulfadimidine

can be given orally or parenterally. Sulfamethoxazole (which was commonly employed with trimethoprim in the combination cotrimoxazole) is well absorbed from the gastrointestinal tract. Phthalyl and succinyl sulfathiazole are poorly absorbed, and can be used to 'sterilize' the gut prior to surgery. Sodium sulfacetamide is not an irritant when applied to the eye, and penetrates well into the ocular fluid. Silver sulfadiazine is used topically on burns, as it is bactericidal to *P. aeruginosa*.

2.6.2 Distribution and duration of action

Following absorption, sulfonamides are transported either in physical solution in the plasma, or bound to plasma proteins. At therapeutically effective doses, sufficient drug is free in the plasma for effective concentrations to build up in the target tissues. Sulfonamides enter most of the body compartments, including the cerebrospinal fluid. Provided that an adequate blood supply is available to a tissue, the concentration in that tissue may be 50–80% of the unbound concentration in the plasma. The permeability of the meninges to sulfonamides is increased when they are inflamed.

2.6.3 Metabolism

Sulfonamides are acetylated in the liver, with acetyl coenzyme A as the acetyl donor, to form less active derivatives. In the event of liver damage acetylation may be reduced, so that the half-life of the drug in the plasma is increased and the frequency of administration should be reduced. The ability to acetylate differs, but whether a patient is a fast or slow acetylator is not significant in sulfonamide chemotherapy. The toxicity of sulfonamides is not lost on metabolism, indeed it may be increased.

2.6.4 Excretion

The rate of excretion of sulfonamides and their metabolites depends on their properties and on kidney function. In kidney failure they may accumulate and doses must be reduced, or sulfonamides avoided when kidney function is impaired. The risk of formation of crystals in the urine (crystalluria) is reduced with highly water-soluble sulfonamides. Adequate water intake should be maintained, as even highly water-soluble sulfonamides may crystallize if the urine flow is small. Making the urine alkaline by giving sodium bicarbonate favors ionization and excretion and reduces renal damage. Acetylation of sulfonamides increases their acidity, and the derivatives are more highly ionized at physiological pH values. Mixtures of sulfonamides are used in order to decrease the risk of precipitation of crystals in the urine, because the solubility of each is independent of the concentration of others, whereas the chemotherapeutic effect is the sum of those of the constituents.

2.6.5 Undesirable effects of sulfonamides

Although many sulfonamides are absorbed through the skin, they are liable to cause skin sensitization and rashes. Skin rashes, itching and irritation of mucous membranes may occur in sensitive individuals, and the incidence of such reactions increases with higher doses. Adverse reactions to sulfonamides occur in about 5% of the population, but minor side-effects may indicate that a more serious reaction may follow, and implies that other sulfonamides will also be toxic, so a history of sensitivity contraindicates their use. Rapidly excreted sulfonamides are preferred when the sensitivity of an individual is not known, so that the duration of any side-effects will be short.

Fever may be induced after about 1 week of administration, so treatment must be closely supervised. Hypersensitivity reactions include impairment of kidney function, and renal function should be monitored during sulfonamide chemotherapy. Sulfonamides rarely affect the blood cells, but such effects may be unrelated to drug dose and may occur after the treatment has been completed. Sulfadiazine causes acute hemolytic anemia and agranulocytosis more frequently than other sulfonamides. Aplastic anemia, thrombocytopenia and eosinophilia may also be caused by sulfonamides.

2.6.6 Interactions of sulfonamides

Some sulfonamides have a high affinity for the plasma-protein binding sites to which tolbutamide and bilirubin also bind. The administration of a sulfonamide with tolbutamide causes an earlier, greater and more long-lasting hypoglycemia than when tolbutamide is used alone. In the neonate, sulfonamides can cause jaundice by displacing bilirubin from the plasma proteins. An increase in plasma bilirubin in the neonate (kernicterus) may occur if sulfonamides are given during pregnancy.

2.7 Other folic acid antimetabolites

2.7.1 p-Amino salicylate

The search for useful chemotherapeutic agents has involved the synthesis of other competitors of p-ABA. The growth of the tubercle bacillus *Mycobacterium tuberculosis* was inhibited by p-aminosalicylic acid (pAS) (*Figure 2.5*), and most strains are sensitive *in vitro* to concentrations as low as 1 µg ml^{-1}. The drug is thought to inhibit folic acid

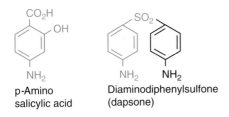

Figure 2.5: pAS and dapsone, congeners of sulfonamide effective against mycobacterium species.

synthesis as an analog of pABA. The specificity implies an ability of the target enzyme to bind pAS that is not found in other micro-organisms. Analogs of folate are formed from the pAS used in the treatment of *M. tuberculosis*. Chemotherapy of tuberculosis requires 8–12 g daily of pAS, or less if it is combined with other drugs such as isoniazid (isonicotinic acid hydrazide). The advent of antibiotics that are effective against mycobacteria (streptomycin, rifampicin) has reduced the use of pAS.

2.7.2 Dapsone

By contrast with pAS and tuberculosis, the sulfone drug dapsone (4:4' diaminodiphenylsulfone) (*Figure 2.5*) remains useful for treating leprosy. Although more effective against streptococcal infections than sulfanilamide, dapsone is also more toxic and was considered unlikely to be useful clinically. However, dapsone was able to suppress experimental tuberculosis infections, and most strains of *M. tuberculosis* are inhibited by 10 μg ml^{-1} *in vitro*. Clinical trials of the use of dapsone in leprosy were conducted in the 1940s, but investigation of the mode of action was hampered because mycobacteria are difficult to investigate *in vitro*. The mode of action is presumed to be similar to that of the sulfonamides and pAS. The derivative 4:4'-diacetyldapsone (which has also been used as a repository antimalarial drug) has a longer duration of action in the treatment of leprosy than dapsone.

2.8 Dihydrofolate reductase inhibitors

The enzyme DHFR reduces dihydrofolate to tetrahydrofolate, the active form of folic acid (*Figure 2.3*), which is essential to all cells, whether they manufacture folic acid (sulfonamide-sensitive bacteria) or require pre-synthesized folic acid (vitamin-requiring higher organisms). Analogs that closely resemble folic acid (e.g. methotrexate) are taken up by mammalian cells and inhibit the mammalian DHFR. Such folate analogs are not absorbed by bacteria and are therefore of no use in the

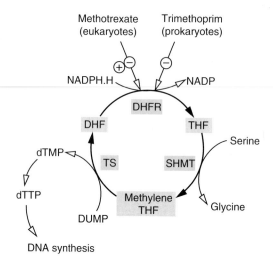

Figure 2.6: The thymidylate cycle and dihydrofolate reductase inhibitors. DHFR = dihydrofolate reductase; SHMT = serine hydroxy methyl transferase; TS = thymidylate synthase.

chemotherapy of bacterial infections. Dihydrofolate reductases from different organisms do not have identical properties, and surveys of a wide range of compounds have resulted in the identification of drugs that inhibit the enzyme in protozoa and in bacteria, but not in higher organisms [6, 7].

The methylation that forms the thymine base of thymidine (*Figure 2.6*) takes place at the level of the nucleoside monophosphate, deoxythymidine monophosphate (dTMP), and is significant because the reaction converts methylene tetrahydrofolate to the oxidized dihydrofolate. The formation of thymidine phosphates for DNA synthesis therefore depletes the tetrahydrofolate pool, and the activity of DHFR is essential for DNA synthesis and cell growth.

Close analogs of folates, such as methotrexate (*Figure 2.7*), are inhibitors of this enzyme in higher organisms and are not applicable to bacterial infections, but are used to treat leukemias in children.

2.9 Trimethoprim and other diaminopyrimidines

The 2:4 diaminopyrimidines are a series of drugs identified by screening for compounds that selectively inhibit bacterial DHFR [6, 8]. The affinity of mammalian DHFR is so low that the concentrations of drug that inhibit micro-organisms have little effect on the host. Pyrimethamine is active against plasmodia (Section 7.1.1) and against bacteria, although other congeners are more effective. The concentration required to

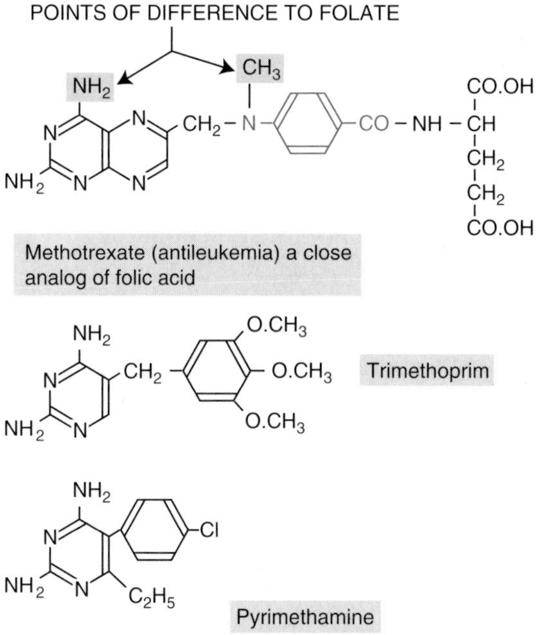

POINTS OF DIFFERENCE TO FOLATE

Methotrexate (antileukemia) a close
analog of folic acid

Trimethoprim

Pyrimethamine

Figure 2.7: Methotrexate and the diaminopyrimidines as DHFR inhibitors.

inhibit the parasite enzyme is 10^3–10^5 times less than that required to inhibit mammalian enzymes. Trimethoprim is a pteridine analog that binds to the bacterial DHFR, preventing the conversion of DHF into the useful form, and is more than 10^4 times more effective against bacterial DHFR than against the mammalian enzyme [8]. The inhibition achieved depends on the nutrients available, and thymidine specifically reverses the effects. The result of reductase inhibition is the cessation of synthesis of purines, pyrimidines, methionine, glycine, histidine, pantothenic acid and N-formylmethionyl tRNA.

2.9.1 Cotrimoxazole

The synergism between sulfonamides and DHFR inhibitors *in vitro* presumably occurs because they affect the same pathway at different sites (*Figure 2.4*). The synergistic effects of two drugs in combination have been illustrated by an isobologram, and can be seen on plates used for testing susceptibility by diffusion of the drugs from paper discs. Trimethoprim was originally designated a 'sulfonamide potentiator', although it is increasingly used alone. It was proposed that the trimethoprim–sulfamethoxazole (cotrimoxazole) combination was less susceptible to resistance. Because of the rapid onset of activity of trimethoprim, the combination is reported to be more immediately effective than sulfonamides alone. The bactericidal spectrum of cotrimoxazole is also wider than that of either of its constituents, because the combination is able to stop growth of strains against which each constituent is only marginally active. Doubts have been expressed about whether synergy occurs under clinical conditions, and about the toxicity of the sulfonamide in the combination [9]. High levels of resistance to cotrimoxazole have been reported in some developing countries [10]. Single-dose therapy with trimethoprim or cotrimoxazole [11] is effective against infections of the bladder, but not of the kidney.

Cotrimoxazole is effective against *C. diphtheriae, Streptococcus pneumoniae* and *N. meningitidis* as well as many enteric microorganisms, and was often used to treat urinary tract infections. In many applications the combination is only used if it is clear that the infective agent is sensitive and also if there is no satisfactory alternative. Methicillin-resistant *S. aureus* is often sensitive to the combination of cotrimoxazole, but not necessarily to either constituent separately. Despite the synergy between the components, for most purposes cotrimoxazole has been replaced by trimethoprim alone, but the combination is still recommended for treatment of *Pneumocystis carinii* pneumonia.

2.9.2 Fansidar, proguanil and trimetrexate in eukaryotes

The Fansidar combination includes the effective inhibitor of protozoal DHFR, pyrimethamine with sulfadoxidine, and is used for the prophylaxis and treatment of malaria (see *Figure 7.2*). Paludrine (chlorguanide, proguanil), which has also been used in the prophylaxis of malaria [12], is metabolized in the liver to a substance that resembles pyrimethamine and is also a DHFR inhibitor in plasmodia (Section 7.1.1). Trimetrexate has been used as an antitumor agent, but is also used to treat the fungus *P. carinii* [13] in AIDS patients. In these applications the host tissues are protected by the simultaneous administration of leucovorin (folinic acid), which cannot be transported into the fungus or, presumably, some tumors.

2.10 Pharmacology of the DHFR inhibitors

Trimethoprim is largely excreted unmetabolized in the urine, and because it is a weak base, excretion is particularly rapid if the urine is acidic. The side-effects of trimethoprim are similar to those of sulfonamides, and disturbances of folate metabolism that can be attributed unequivocally to trimethoprim are blood disorders such as reduced erythropoiesis, leukopenia and thrombocytopenia. Normal blood-cell development can be re-established by giving folinic acid (formyl-THF) to patients who show these side-effects. Cotrimoxazole is well absorbed from the gastrointestinal tract, and the combination is reported to give the optimum synergistic plasma levels of the two compounds.

References

1. Huennekens, F.M., Vitols, K.S. and Henderson, G.B. (1978) *Adv. Enzymol.*, **47**, 313–346.
2. Woods, D.D. (1962) *J. Gen. Microbiol.*, **29**, 687–702.
3. Wolf, B. and Hotchkiss, R.D. (1963). *Biochemistry*, **2**, 145–150.
4. Brown, G.M. (1962) *J. Biol. Chem.*, **237**, 536.
5. Swedberg, G., Castensson, S. and Skold, O. (1979) *J. Bacteriol.*, **137**, 129–136.
6. Schweitzer, B.I., Dicker, A.P. and Bertino, J.R. (1990) *FASEB J.*, **4**, 2441–2452.
7. Gready, J.E. (1980) *Adv. Pharmacol. Chemother.*, **17**, 37–102.
8. Baker, D.J., Beddell, G.R. and Champress, J.N. (1981) *FEBS Lett.*, **126**, 49–52.
9. Ball, P. (1986) *J. Antimicrob. Chemother.*, **17**, 694–696.
10. Murray, B.E., Alvaradet, T. and Kim, K.H. (1985) *J. Infect. Dis.*, **152**, 1107–1113.
11. Ronald, A.R., Conway, B. and Zhanel, G.G. (1990) *Chemotherapy*, **36**, (Suppl.), 2–9.
12. Peterson, D.S., Milhous, W.K. and Wellems, T.E. (1990) *Proc. Natl Acad. Sci. USA*, **87**, 3018–3022.
13. Allegra, C.J., Chabner, B.A., Tuazon, C.U. *et al.* (1987) *N. Engl. J. Med.*, **317**, 978–985.

Chapter 3

The cell wall as a target for antimicrobial drugs

3.1 History and introduction to antibiotics active at cell walls

In 1928, Fleming noticed that staphylococci were lysed (broken open) on plates infected with a mold of the genus *Penicillium*. The culture medium in which the mold was grown was bactericidal, although not to Gram-negative bacteria such as *E. coli*, yet it was not toxic to humans or animals. The name 'penicillin' was given to filtrates of the mold culture, but inability to isolate the active ingredient hindered progress until Chain and Florey purified a number of penicillins. The development of culture techniques and high-yielding mutants subsequently made the production of pure penicillins possible.

Sublethal concentrations of penicillin cause lengthening and swelling of sensitive bacteria. It was proposed that penicillin interfered with normal cell-wall formation to produce weak walls that might burst with consequent loss of the cell contents. In 1949 Park described the accumulation of nucleotides containing amino acids in staphylococci treated with penicillin, and these compounds were identified as precursors of the peptidoglycan layer (also called mucopeptide) of the cell wall. The structure and synthesis of this important component have been elucidated in parallel with the identification of the site of action of several antibiotics. It is now known that β-lactams react with several binding proteins (PBPs) in the bacterial membrane, which are enzymes involved in the synthesis of peptidoglycan.

3.2 The structure of peptidoglycan in the bacterial cell wall

The membrane of bacteria surrounds a cytoplasm with relatively high concentrations of low-molecular-weight substrates, which therefore has

Table 3.1: Components of bacterial cell walls

	Gram-positive	Gram-negative
Peptidoglycan	+	+
Teichoic and teichuronic acids	+	–
Lipoprotein	–	+
Lipopolysaccharide	–	+
Protein	Some strains	+
Polysaccharide	+	+

(see *Figure 6.1*).

a high osmotic pressure. Bacteria have an exterior cell wall against which the cell membrane is pressed by turgor pressure. The major structural component of the wall, from the standpoint of mechanical strength, is the peptidoglycan (*Table 3.1*). Treatment with lytic enzymes (lysozyme) that hydrolyze peptidoglycan causes loss of cell shape and integrity.

Gram-positive bacteria, which stain purple, have cell walls that are 200–500 Å thick and are relatively simple in structure. A layer of peptidoglycan next to the cell membrane contributes about 50% of the total mass of the walls (*Table 3.1*). Gram-negative bacteria stain pink, due to the counterstain. The cytoplasm contains lower concentrations of metabolites, so the internal osmotic pressure may be as low as five atmospheres compared to Gram-positive bacteria. The cell wall is thin (100–150 Å) but very complex, with more than one layer. The inner layer is composed of peptidoglycan and may represent as little as 5–10% of the cell wall. Despite constituting a low proportion of the wall, peptidoglycan still determines the shape of the cells and confers rigidity and strength on the walls.

Failure to synthesize adequate amounts of peptidoglycan has different results in different bacteria. Typically, growing bacteria swell and lyse due to their internal osmotic pressure. This effect is evident at the sites where new wall material is intercalated into the existing structure, and cross-linked. In bacilli, aberrant morphological forms (e.g. 'rabbit-ear' shapes) may be seen. The ends of the rods, comprising 'old' peptidoglycan synthesized before treatment, retain their shape (the ears). The cells balloon out in regions where new synthesis of weakened peptidoglycan is occurring. These antibiotics are therefore bactericidal towards growing cells. Drugs that inhibit peptidoglycan synthesis have no effect on the mature peptidoglycan of nongrowing bacteria, which may survive.

3.3   The peptidoglycan target of action of antibiotics

Peptidoglycan is a lattice-like macromolecule (*Figure 3.1*) composed of polysaccharide (glycan) strands oriented parallel to one another and cross-linked by short peptides. The glycan chains vary little from one bacterium to another (*Figure 3.2*), and consist of N-acetylglucosamine

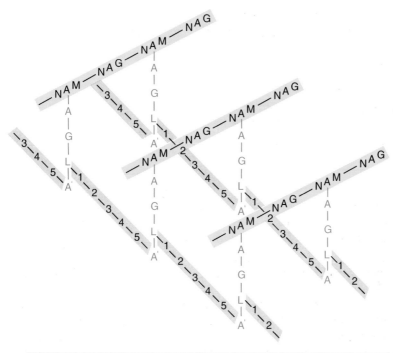

Carbohydrate strands consisting of alternate N-acetylmuramic acid (NAM) and N-acetyl glucosamine (NAG) are linked via tetrapeptides (A-G-L-A') cross-linked by pentaglycine chains (1-2-3-4-5).

Figure 3.1: The structure of peptidoglycan (mucopeptide) of *Staphylococcus aureus.*

(NAG) alternating with the 3-O-lactyl ether of N-acetylglucosamine, namely N-acetylmuramic acid (NAM), linked together by β1:4 glyco-side bonds. Peptides are attached to the glycan by peptide bonds between the amino group of the first amino acid in the chain and the carboxyl group of the lactic acid residue of NAM. The peptide is peculiar in that it contains alternating D and L amino acids, and in mature peptidoglycan NAM is linked to a tetrapeptide (A-G-L-A') (*Figure 3.1*) in which A is L-alanine, G is D-glutamic acid or D-glutamine, L is L-lysine or another dibasic amino acid and A' is D-alanine. In this struc-ture the glutamic acid or glutamine residue (*Figure 3.2*) is linked in a peptide bond via its 5-carboxyl rather than via its 1-carboxyl, as it is in proteins. This unusual arrangement is distinguished by describing the residues as D-isoglutamic acid or D-isoglutamine. The tetrapeptides are cross-linked from the side-chain amino group of the dibasic amino acid (L) to the terminal carboxyl (of residue A) of another tetrapeptide chain. The cross-link is between tetrapeptides attached to different

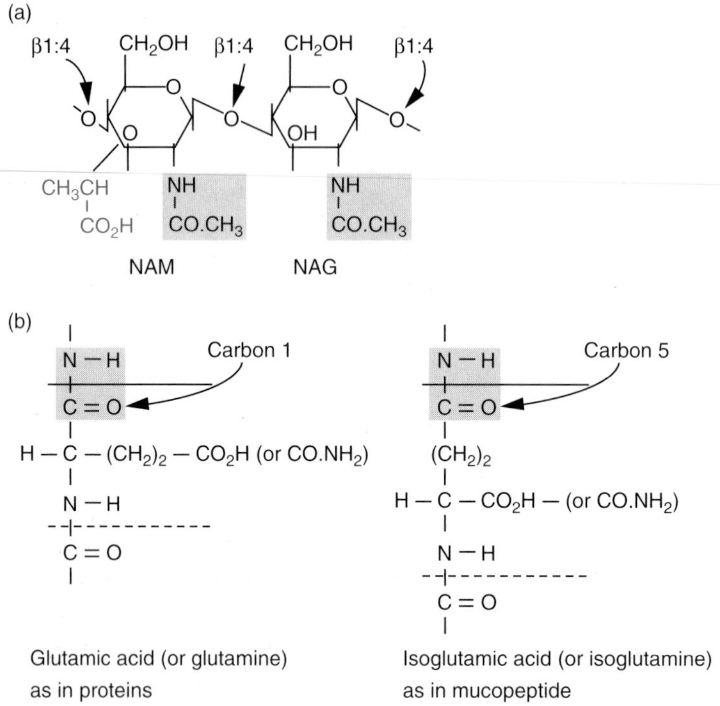

(a)

NAM   NAG

(b)

Glutamic acid (or glutamine)   Isoglutamic acid (or isoglutamine)
as in proteins   as in mucopeptide

Figure 3.2: Structural features of peptidoglycan: (a) The β1:4-linked glycan chain of N-acetylmuramic acid (NAM) and N-acetylglucosamine (NAG). (b) Glutamate or glutamine linked in a peptide bond via C5 rather than via C1 as in proteins.

glycan chains, producing a network (*Figure 3.1*). In many cases the cross-link between the peptides is direct, but in some instances bridges are composed of other amino acids. The peptidoglycan of *S. aureus* has lysine linked to D-alanine by a pentapeptide (1-2-3-4-5 in *Figure 3.1*) composed of glycine residues.

There are considerable variations in the structure of peptidoglycan in different bacteria although there is very little difference in the glycan. In particular a number of alternative diamino acids occur in place of lysine, and there are considerable variations in the detail of the cross-linking, which may be direct or via oligopeptides.

When *S. aureus* is treated with sublethal concentrations of penicillin, compounds accumulate that contain a nucleotide, amino sugars and several amino acids, which are the precursors of the peptidoglycan layer. Cell walls synthesized in the presence of penicillin contain more alanine than normal, and the glycine residues have free amino groups. This indicates that failure to cross-link the peptidoglycan forms the basis for the action of penicillin, and is consistent with the effects of penicillin on cell morphology.

3.3.1 The synthesis of peptidoglycan

Peptidoglycan is made in several stages. The initial reactions occur in the cytosol of the cell (*Figure 3.3*), further transformations are then effected in the cell membrane (*Figure 3.4*), and the final incorporation of peptidoglycan into the bacterial cell wall occurs at the points of extension (intercalation) of the cell wall. The enzymes involved in this phase have been identified with some penicillin binding proteins (PBPs). Control of the action of the various PBPs is responsible for determining the cell shape, increasing the size of the peptidoglycan envelope during growth and the formation of septa for cell division.

The cytosolic phase of synthesis. The basic subunit of peptidoglycan made in the cytoplasm consists of NAM attached to the activating nucleotide uridine diphosphate (UDP) and bearing a pentapeptide (A-G-L-A'-A'), not a tetrapeptide (A-G-L-A') as in the mature peptidoglycan (*Figure 3.1*). N-acetylglucosamine-1-phosphate is first converted into UDP-NAG (*Figure 3.3*). The lactic acid residue that converts

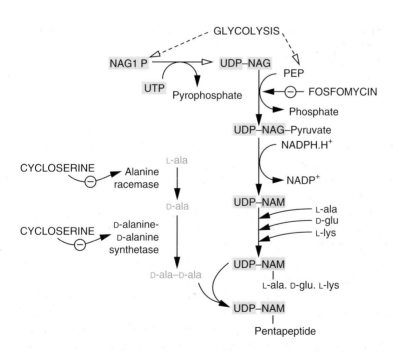

Figure 3.3: The assembly of peptidoglycan precursor in the bacterial cytoplasm.

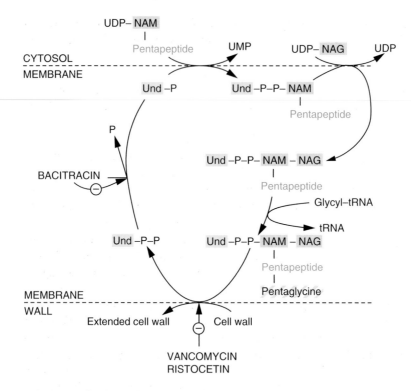

Figure 3.4: The synthesis and transfer across the membrane of the complete peptidoglycan subunit.

UDP-NAG into UDP-NAM is incorporated as pyruvate from the glycolytic intermediate phosphoenolpyruvate (PEP), followed by reduction of this pyruvate to lactate. The lactate is joined by an ether link between its hydroxyl and the 3-hydroxyl of the NAG. The peptide is attached to the carboxyl group of the lactic acid residue. The stepwise addition of three amino acids by peptide bonds forms the tripeptide derivative UDP-NAM-tripeptide. The last two amino acids of the pentapeptide are added as a dipeptide of D-alanine, which is synthesized separately by D-ala-D-ala synthetase. The D-alanine is produced from L-alanine by alanine racemase.

The membrane phase of synthesis. The next stage of synthesis takes place in the cell membrane (*Figure 3.4*). The membrane carrier used is the 55-carbon (C55) lipid undecaprenyl phosphate (Und-P), comprising 11 isoprene units. This lipid accepts phospho-NAM-pentapeptide from UDP-NAM-pentapeptide in a reaction involving translocase I, releasing UMP into the cytosol. Some UDP-NAM-tripeptide is also incorporated from the cytosol on to Und-P. These lipid complexes in the membrane then accept NAG from cytoplasmic UDP-NAG in a transfer using trans-locase II, so that the growing peptidoglycan subunit now contains the NAM-β1:4-NAG disaccharide subunit and either tri- or pentapeptide. In the case of *S. aureus*, the pentaglycine side-chain is combined with the lysine amino group (of A-G-L-A'-A'), the donor being a specific glycyl-tRNA that is distinct from that involved in protein biosynthesis (*Figure 3.4*).

The cell-wall phase of synthesis. Growing bacteria have lytic enzymes to hydrolyze the mucopeptide locally in order to allow new components to be added (intercalated) at growing points. The peptidoglycan subunit in the membrane is detached from the undecaprenyl pyrophosphate (Und-P-P) carrier and transferred to a growing point in the peptido-glycan by a bond to the NAM-NAG disaccharide. The released un-decaprenyl lipid bears pyrophosphate, which must be hydrolyzed by a specific pyrophosphatase to form the monophosphate that may again accept UDP-NAM-pentapeptide from the cytosol. If peptidoglycan were synthesized only by the processes described so far, no lateral cross-links between the glycan chains would be established. In the cross-linkage of peptidoglycan by transpeptidation (*Figure 3.5*), the side-chain amino group of the pentaglycine of one glycan chain reacts enzymatically with the peptide bond between two D-alanine residues of a pentapeptide (described above as A-G-L-A'-A') from another glycan chain. The reac-tion involves the migration of the peptide bond and the transfer of a proton from the pentaglycine amino group; free D-alanine is released. Mechanistically, the process proceeds via an acyl-enzyme intermediate between the serine hydroxyl in the enzyme and the carbonyl of the

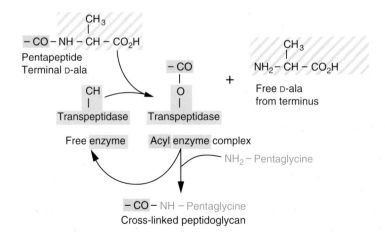

Figure 3.5: Normal transpeptidation reaction.

β-lactam, and it is central to the efficiency of cross-linking that this intermediate is not readily hydrolyzed. The details of the reaction, and the extent of cross-linkage, vary between micro-organisms.

3.4 Inhibitors of the early stages of peptidoglycan synthesis

The processes of peptidoglycan biosynthesis that are affected by antibiotics have been indicated above and have been reviewed frequently [1–7].

3.4.1 Inhibitors of the cytosolic phase of synthesis

Phosphonomycin (fosfomycin) is an antibiotic that has not been used widely in clinical medicine, although it is effective against Gram-negative rods and Gram-positive cocci, and is also nontoxic. Although its structure resembles that of PEP, the resemblance is not close (*Figure 3.6*), and fosfomycin is not a general inhibitor of reactions involving PEP. The transfer of pyruvate from PEP to NAG during synthesis is specifically inhibited by fosfomycin [8–10], but the affinity of the antibiotic for the pyruvate transferase enzyme is low, and high concentrations of antibiotic have to be used. Fosfomycin reacts with a cysteine residue in the enzyme, and the specificity of this reaction contributes to the low toxicity of the drug. The sodium salt may be used parenterally, but for oral treatment of cystitis the trometamol salt is used. D-Cycloserine (oxamycin) is accumulated by the D-alanine transporter of bacteria. The molecule is an analog of D-alanine and inhibits both the racemization of L-alanine and the synthesis of the D-alanine dipeptide [11–13]. Cycloserine is used clinically as a secondary drug for the treatment of *M. tuberculosis*. It is not a drug of first choice because

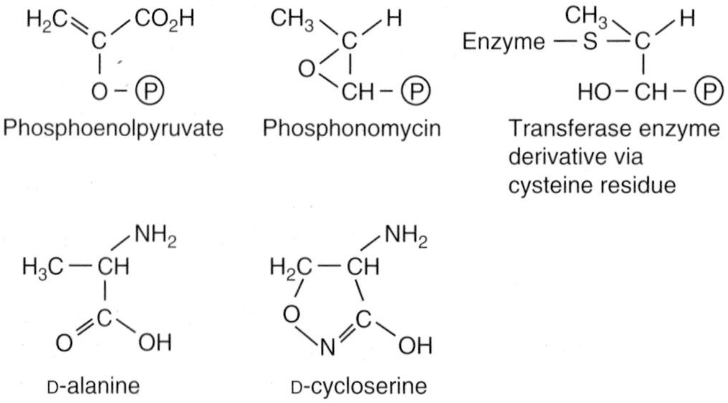

Figure 3.6: Structures of phosphonomycin and cycloserine.

of its deleterious effects on the nervous system. In some bacteria, cycloserine causes the accumulation of a 'Park nucleotide', UDP-NAM-tripeptide, that lacks the usual terminal D-alanine-D-alanine. Inhibition by cycloserine is reduced by exogenous alanine in some strains.

3.4.2 Resistance to fosfomycin

Resistance to fosfomycin may be due to drug modification, or to loss of transport. Seven varieties of plasmids isolated from enterobacteria [14] all carry the same fos^R determinant for a modifying enzyme. A gene, known as *fosB*, on small plasmids (2.4 to 4.1 kb) codes for resistance in staphylococci [15], and nucleotide sequences related to *fosB* occur in five species including *S. aureus* from Australia. The *fosB* gene has also been detected in France and Japan. Weak hybridization with plasmids of staphylococci may indicate resistance heterogeneity in this genus. The spread of fosfomycin resistance seems to be low, despite the diversity of fos^R plasmids.

3.4.3 Inhibitors of the membrane phase of synthesis

Bacitracin is a mixture of three similar peptides (A, B and C) containing linear and cyclic components (*Figure 3.7*). It is too toxic for systemic use, but may be employed topically for skin infections, and is sometimes used to suppress Gram-positive bacteria in abdominal surgery. It binds to Und-P-P and inhibits the membrane pyrophosphatase that releases undecaprenyl phosphate from the pyrophosphate [16, 17]. It therefore causes the accumulation of the lipid carrier in the pyrophosphate form, so preventing the transfer of peptidoglycan precursors from UDP derivatives in the cytoplasm on to undecaprenyl monophosphate in the membrane.

Gram-negative bacteria, with the exception of *N. gonorrhoea*, are not susceptible to vancomycin, which does not enter their cells, but the drug is active against staphylococci, streptococci and other Gram-positive bacteria, including methicillin-resistant *Staphylococcus aureus* (MRSA). Vancomycin blocks the final stage of assembly of peptidoglycan, namely the transfer from the membrane to the growing point of the wall [18]. It does not inhibit the enzyme, but binds to D-ala-D-ala in the undecaprenyl peptidoglycan precursor. Other compounds that contain the dipeptide, including UDP-NAM-pentapeptide, also bind to vancomycin *in vitro*, but this cytoplasmic precursor does not bind *in vivo* because vancomycin does not pass through the cell membrane. The binding of vancomycin to the D-ala-D-ala at the membrane surface prevents the transfer of the subunit into the growing cell wall. Modification of vancomycin near the N-terminus, and particularly the removal of the

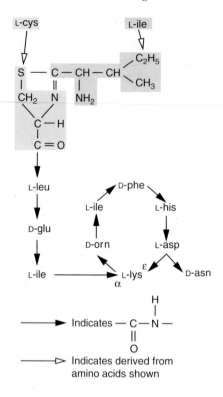

Figure 3.7: Structure of bacitracin A.

end leucine residue, reduces or abolishes activity [19]. Vancomycin is bactericidal in infections with staphylococci and streptococci, including endocarditis, when allergy precludes the use of β-lactams, and resistance has so far not become a major problem despite widespread use of the drug. It is also used to control *C. difficile* superinfections. The toxic effects of vancomycin involve the ear and the kidney.

Ristocetin is a glycopeptide with a similar mechanism of action to vancomycin. The toxicity of both of these drugs precludes their use except for severe infections that fail to respond to other antibiotics. When serum concentrations are monitored and administration is not too rapid, vancomycin toxicity can be minimized, although this problem is said to be less than with early less pure preparations. Teicoplanin is a newer glycopeptide antibiotic with a chemical resemblance to vanco-mycin [20], but with a different spectrum of antibacterial activity, both *in vitro* and *in vivo*. It is now administered in larger doses than was first recommended, and its proposed advantages include less frequent side-effects, lower toxicity (especially when combined with an amino-glycoside), the possibility of intramuscular administration and a low dose frequency.

3.4.4 *Resistance to membrane-active antibiotics*

Vancomycin is the best drug for treatment of MRSA, *Corynebacterium jeikeium*, and *S. pneumoniae* showing multiple drug resistance. The era of vancomycin resistance has been anticipated with foreboding because of its importance in treating these antibiotic-resistant strains.

Enterococci are important nosocomial pathogens among which resistance to multiple antibiotics is occurring with increasing frequency. Glycopeptide resistance in enterococci is of recent origin, and it narrows the options for treatment of infections with these opportunistic pathogens. Low-level determinants of resistance to vancomycin are not transferable and are presumably chromosomal. Resistance to high levels of vancomycin and teicoplanin is mediated by a plasmid determinant distinct from that which confers a low resistance *(Table 3.2)*.

The 38-kDa protein VanA resistance protein of *Enterococcus faecium* is homologous with Gram-negative D-ala-D-ala ligases, the enzymes that catalyze the synthesis of the terminus of peptidoglycan precursors, but with a modified specificity [21]. VanA preferentially condenses D-met or D-phe instead of D-ala, raising the possibility that it synthesizes a modified cell-wall component that is not recognized by vancomycin. Resistance may be due to the combination of D-hydroxybutyrate with D-alanine, catalyzed by a mutant D-ala-D-ala ligase [21]. A cytoplasmic precursor of peptidoglycan that terminates in D-lactate rather than D-alanine, has also been discovered in *Enterococcus faecalis* [22]. The moderate vancomycin resistance (minimum inhibitory concentration (MIC) = μg ml^{-1}) of *Enterococcus gallinarum* is constitutive, but strains are susceptible to the other glycopeptides, like teicoplanin.

Enterococci that are resistant to glycopeptides nevertheless show a synergy between glycopeptides and β-lactams [23]. By contrast, susceptible enterococci are not synergistically inhibited by penicillin and vancomycin. *E. gallinarum* resistance to low levels of vancomycin is due to the gene *vanC* that encodes a mutant D-ala-D-ala ligase, although with only 29–38% identity with VanA and the *E. coli* ligase [24].

Table 3.2: Vancomycin resistance genes in enterococci

	vanA	vanB	vanC
E. faecium	+	+	−
E. faecalis	+	+	−
E. gallinarium	−	−	+
MIC vancomycin (μg ml^{-1})	64	16–32	8–16
MIC teicoplanin (μg ml^{-1})	16	0.5	0.5
Induced by vancomycin	+	+	−
Induced by teicoplanin	+	−	−

The *vanH* gene specifies a keto-acid reductase that makes a hydroxy acid which VanA then couples into the hydroxy acid analogs referred to above [25]. VanY protein is a carboxypeptidase encoded by a high-level vancomycin resistance gene of a plasmid from *E. faecium* [26]. The vanY gene product cleaves the terminal D-alanine residue of UDP-muramyl-pentapeptide and also releases the terminal D-hydroxy acid from peptides produced by VanA. It thus hydrolyzes both D-ala-D-ala and its analogs.

It is not yet clear whether resistance is as infrequent for teicoplanin as for vancomycin, and the latter remains the glycopeptide of choice for infections with Gram-positive bacteria.

3.5   The β-lactam antibiotics

The β-lactam antibiotics are the most widely used antibacterial agents [27], including the penicillins and cephalosporins.

3.5.1 Entry of β-lactams into bacterial cells

The β-lactam drugs are small enough to pass through the porin channels of the outer wall of Gram-negative bacteria. In *S. typhimurium* 90% of entry is by porin channels, and when porins are deficient this contributes to the resistance to some β-lactams. Under the conditions that prevail in the gut, *E. coli* predominantly synthesizes the OmpC porin, but PhoE porin is also produced when inorganic phosphate is growth-limiting. The preference of the major porins for cations is consistent with the low penetration of the negatively charged β-lactams. The susceptibility of gonococci to penicillin G may be related to the preference of the gono-coccal porin for anions.

Pseudomonas aeruginosa lacks the classical porins found in enteric bacteria, which may explain the high resistance of this organism to β-lactams. However, the relationship between the external concentration of β-lactam drugs and their concentration in the periplasmic space of Gram-negative bacteria is not simple, and the permeation of some drugs differs from that predicted.

3.5.2 The penicillins

Penicillins [27] are acyl (R.CO-) derivatives of the penicillin nucleus (6-aminopenicillanic acid), which consists of cysteine and valine condensed together to form a rigid double-ring system (*Figure 3.8*). The molecule resembles the sequence of the terminal dipeptide of uncross-linked mucopeptide, D-alanine-D-alanine, the natural substrate for the cross-linking enzyme transpeptidase (*Figure 3.5*). In β-lactam drugs the

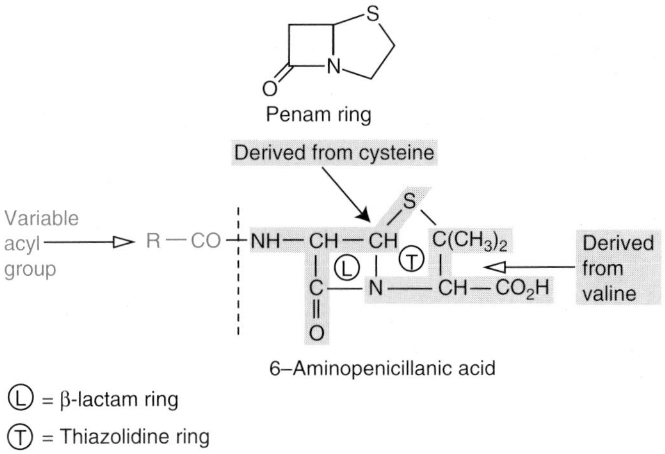

Penam ring

Derived from cysteine

Variable acyl group ⟶ R—CO

Derived from valine

6–Aminopenicillanic acid

Ⓛ = β-lactam ring

Ⓣ = Thiazolidine ring

Figure 3.8: Penicillin structure.

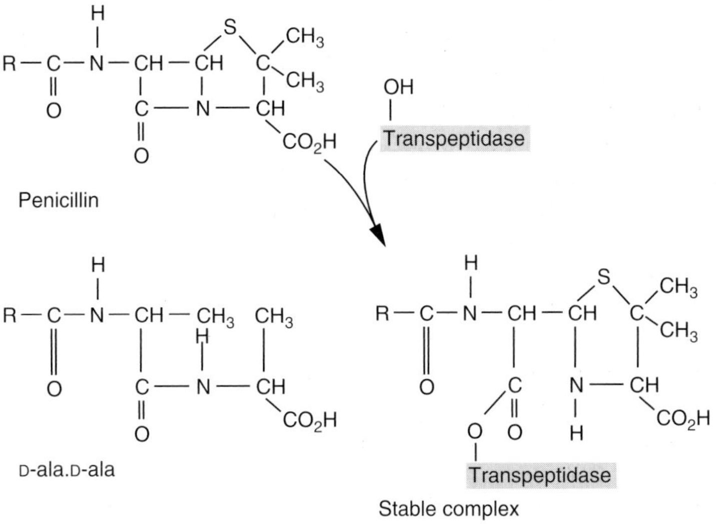

Penicillin

Transpeptidase

D-ala.D-ala

Transpeptidase

Stable complex

Figure 3.9: The transpeptidase reaction with penicillin, an analog of the natural substrate, the D-ala-D-ala of the pentapeptide.

-CO-N- bond of the β-lactam ring is the analog of the peptide bond between the two alanine residues of the natural substrate. During transpeptidation a serine hydroxyl group in the enzyme reacts with alanine-alanine, displacing free alanine and forming an acyl bond between the serine and the penultimate alanine. The enzyme is in turn displaced from this complex by the amino group of the cross-linking

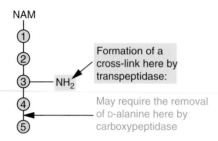

Figure 3.10: Carboxypeptidase action on pentapeptide.

residue. Penicillin and cephalosporin are effective against growing cells, often causing lysis unless the medium and cells are iso-osmotic. A hypothesis thus emerged of penicillin as a substrate analog that reacts with transpeptidase (*Figure 3.9*) to form a stable acyl intermediate. The β-lactam ring acylates the hydroxyl group of one specific serine residue in the enzyme, producing an inactive penicilloyl–enzyme complex. A second enzyme, carboxypeptidase, is analogous to transpeptidase but hydrolyzes the D-ala-D-ala termini of pentapeptides, releasing free D-alanine without cross-linking. The role of this enzyme seems to vary in different micro-organisms. It may compete with transpeptidase in order to limit the extent of cross-linkage and prevent excessive rigidity. In some strains, the action of carboxypeptidase upon a pentapeptide may be a prerequisite for the cross-linking reaction involving the amino side-chain of that pentapeptide (*Figure 3.10*).

3.5.3 The cephalosporins

Cephalosporin antibiotics [27] have a β-lactam ring fused with a larger, six membered ring than in the penicillins (*Figure 3.11*). The mechanism of action is identical to that of the penicillins, but the spectrum of microbial sensitivity to cephalosporins is broad and extends to Gram-negative bacteria.

3.5.4 Other β-lactam compounds

A range of diverse β-lactams, that are neither penicillins nor cephalosporins, have been discovered providing hope for further diversity of therapeutic action [28]. The nocardicidins and monobactams have little antibacterial activity, but a monobactam derivative, aztreonam, is resistant to β-lactamases and active towards the aerobic Gram-negative bacteria. Thienamycin [29], a type of lactam drug (*Figure 3.11*) has a hydroxethyl group in place of the normal amide linked to the β-lactam

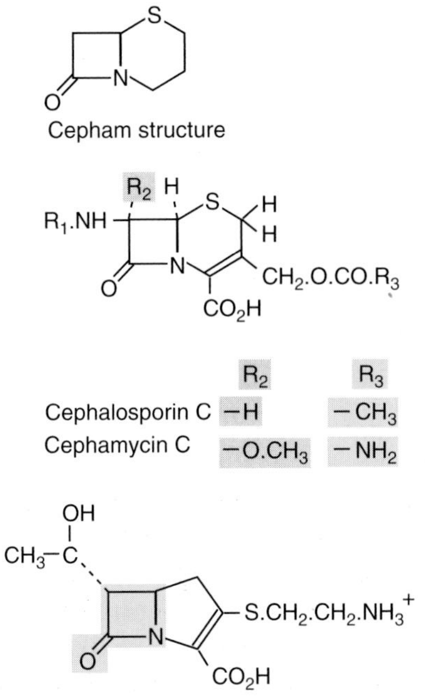

Cepham structure

	R_2	R_3
Cephalosporin C	$-H$	$-CH_3$
Cephamycin C	$-O.CH_3$	$-NH_2$

Figure 3.11: Structures of cephalosporins and thienamycin.

ring, and is stable to lactamases. Thienamycin does not have a thiazo-lidine ring, but does have a sulfur-containing side-chain.

3.5.5 The acyl-enzyme theory of β-lactam action

Transpeptidase activity and the degree of peptidoglycan cross-linking is highest in newly divided cells. High carboxypeptidase activity and a lower degree of cross-linking occur before cell division. Much of the earlier classical chemical inhibition work cites transpeptidase as the site of a single target inhibition producing an inactive acyl-enzyme intermediate [30]. Concerning this, Waxman [31] made the following mechanistic predictions.

(1) The inactive transpeptidase would be a penicilloyl–enzyme complex.
(2) Acyl-enzyme intermediates would occur in the uninhibited reaction.
(3) The acyl group derived from either the substrate or the penicillin moiety would be substituted on the same residue.

Unfortunately, verification of these mechanistic predictions was undertaken with carboxypeptidase enzymes (which can be purified read-ily) rather than with the transpeptidases.

3.5.6 Penicillin binding proteins and β-lactamase inhibitors

The enzymologist's concept of the mechanism of action of the β-lactam drugs, however attractive, was complicated by the diverse effects of β-lactams in some bacteria. The discovery of several PBPs in most bacteria helped to explain these diverse effects. The PBPs were detected by allowing radioactive penicillin to react with membrane proteins, followed by electrophoresis of the labeled proteins in denaturing gels, and autoradiography [32, 33]. This protocol allows the number of PBPs, their sizes and their relative concentrations to be assessed. By conducting competition experiments using two β-lactams, the specificity of the proteins for particular drugs can be deduced.

There are seven PBPs in *E. coli*, numbered from 1 in order of descending molecular size as indicated by their mobility on electrophoresis through denaturing gels (sodium dodecylsulfate–polyacrylamide gel electrophoresis SDS–PAGE); (*Figure 3.12*), and up to five PBPs are detectable in *Bacillus* species. Three enzyme activities (D-alanine carboxypeptidase, peptidoglycan transpeptidase and peptidoglycan endopeptidase) are assigned to these PBPs, all of which react with penicillin G in *E. coli*. The enzyme activities of the PBPs and the effects of inhibiting them are summarized in *Figure 3.12*.

PBP1 has two components, PBP1A and PB1B [3], each of which has both glycan-polymerizing and cross-linking (transpeptidase) activities. PBP1B is the major transpeptidase of *E. coli*, but some effective β-lactam drugs do not inhibit this enzyme. It is not clear whether transpeptidase activity in *E. coli* is solely due to the PBP1 component. Mutants that lack both these enzyme activities have them restored by the addition of pure PBP1B. The binding of PBP1B to cephalosporins and penicillins is followed by cessation of cell elongation and by lysis, so it is proposed that this protein is a requirement for peptidoglycan synthesis. PBP1A does not appear to be essential for normal peptido-

Protein	Gel pattern	Protein M$_r$ (kDa)	Enzyme activity
PBP1A		91	Transpeptidase–transglycosylase
PBP1B		90	Transpeptidase–transglycosylase
PBP2		66	Transpeptidase
PBP3		60	Transpeptidase–transglycosylase (cross wall active)
PBP4		49	Carboxypeptidase
PBP5		42	Carboxypeptidase
PBP6		40	Carboxypeptidase

Figure 3.12: Penicillin binding proteins of *Escherichia coli*.

Table 3.3: Effects of lactams on six classes of β-lactamases

Enzyme	Cephalo-sporin	Penicillin	Isooxacyl penicillin	Carbenicillin	Clavulanate	Sulbactam
I	H[a]	–	I[b]	partI	–	–
II	–	H	I	I	I	I
III	H	H	I	I	I	I
IV	H	H	–	–	I	I
V	H	H	H	–	–	I
VI	H	H	I	I	–	–

[a]Drug hydrolyzed.
[b]Enzyme inhibited.

glycan synthesis. PBP1 has the highest affinity for cephaloridine, which causes lysis before any changes in cell shape are detectable [3, 3].

Some β-lactam drugs are specific and their effects have illuminated the roles of certain PBPs. PBP3 binds almost exclusively to cephalexin [3, 34], and as it also inhibits cell division and causes filamentation, it may be required for cross-septum formation. Peptidoglycan in cephalexin-treated filamentous cells of *E. coli* is highly cross-linked, perhaps due to inhibition of a carboxypeptidase that limits the extent of cross-linking.

Mecillinam (amidinocillin) has the amino group of the penicillin nucleus derivatized by an amidino group rather than being acylated. It is active against ampicillin-resistant enterobacteria, but less widely active against Gram-positive organisms. Mecillinam binds preferentially to PBP2 and it produces ovoid cells without any defect in the peptidoglycan. The shape change induced by inhibition of PBP2 leaves the cells in an osmotically stable state [3]. The β-lactams are therefore effective if they interact with PBP1B, PBP2 or PBP3 [3]. Mecillinam and cephalexin are effective specifically as a result of binding to their respective targets, PBP2 and PBP3.

The binding of β-lactams to PBP4, PBP5 and PBP6, which are carboxypeptidases, appears to be irrelevant to the action of the drugs [3]. The carboxypeptidases also catalyze weak transpeptidase activity *in vitro*, but there is no evidence that they do so *in vivo*. Carboxypeptidases are inhibited by concentrations of β-lactams much lower than those which inhibit growth. Mutants that lack carboxypeptidase 1B activity also lack PBP4, and therefore these entities are identical. Carboxypeptidase 1A has been identified with PBP5 and PBP6, a doublet in SDS gel electrophoresis. PBP4, PBP5 and PBP6 are responsible for a large proportion of penicillin binding, but their inhibition is not lethal, and mutants that have lost their carboxypeptidase activity show no physiological impairment.

Rod-shaped bacteria have one major (or two closely related) PBP that accounts for 70–90% of the penicillin bound, and is a carboxypeptidase.

It has been suggested that this carboxypeptidase acts as a transpeptidase *in vivo*, which is 'uncoupled' during solubilization. It may also act to limit the degree of cross-linkage and allow flexibility.

PBP7 appears to be related to lysis in nongrowing *E.coli* [35]. The PBPs are structurally altered in penicillin-resistant strains.

3.5.7 Autolysins and the mechanism of action of β-lactams

Attachment of penicillin to PBPs only results in inhibition of growth. The irreversible effects of penicillin require peptidoglycan hydrolysis. A role for autolysins (cell-wall-degrading enzymes that act from within, causing the cells to lyse) in penicillin-induced lysis has been proposed [36, 37]. Autolysin-defective mutants of pneumococci lack the enzyme L-alanyl-N-acetylmuramic acid amidase, and they fail to lyse when treated with penicillin (they are said to display 'penicillin tolerance'). The addition of purified exogenous autolysin to such penicillin-treated, autolysin-defective pneumococci causes rapid lysis. Low concentrations of penicillin, sufficient to cause lysis in a wild-type strain, cause leakage of lipoteichoic acid (LTA) (*Figure 3.13*) from the walls of lysis-defective cells. LTA is a powerful inhibitor of the lytic enzyme N-acetylmuramyl-L-alanine amidase [38]. The inhibition of growth, sensitization of the cells to added autolytic enzyme and the leakage of LTA from the cells all occur at penicillin concentrations that cause the wild-type cells to lyse.

Two mechanisms have been proposed for the degradation of peptidoglycan. The 'constitutive' model [1] assumes that autolytic hydrolases and the synthetic enzymes act together, with carefully controlled hydrolase action supplying growing points for the extension of peptidoglycan. The inhibition of the synthetic enzymes alone therefore leads to lysis of growing cells.

The second model (autolysin triggering) was proposed to explain the irreversible effects of penicillin in pneumococci [1]. This model proposes that the primary effects of penicillin on enzymes corresponding to

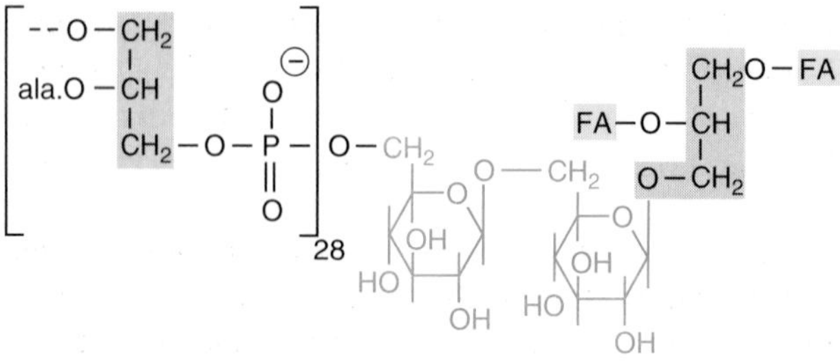

Figure 3.13: Glycerol lipoteichoic acid of *Staphylococcus aureus*.

PBPs, including transpeptidase or carboxypeptidase, lead to cessation of growth. This is followed by the secondary (irreversible) effects of penicillin, the main biochemical events of which are as follows.

(1) Dissociation of the autolysin–inhibitor complex with release of the inhibitor (LTA) from the cells.
(2) Inhibitor loss triggering hydrolase activity and cell lysis.

3.6 Administration, metabolism and excretion of β-lactams

Penicillins are absorbed from the stomach and duodenum, but are destroyed in the large intestine. Penicillin V, ampicillin, cloxacillin and flucloxacillin resist hydrolysis by stomach acid and may be given orally. Those penicillins that are sensitive to acid (e.g. penicillin G, methicillin, carbenicillin) are usually injected intramuscularly, although oral administration is possible if stomach acidity is low, as in the very young and the very old.

When these drugs are injected intramuscularly, both distribution and excretion are rapid, so large or frequent injections must be given. The duration of action from the deep intramuscular injection site is extended by equimolar procaine or benzathine for penicillin G, or by probenecid for methicillin and carbenicillin. Penicillins can also be given intravenously, subcutaneously or intrathecally and penetrate into most tissues, crossing the placental membranes and entering the cerebrospinal fluid (CSF). The binding of penicillins to plasma proteins varies, but sufficient unbound drug remains for activity. The displacement of penicillins from plasma proteins by other drugs does not appear to be significant.

Cephalothin and cephaloridine are poorly absorbed from the gastrointestinal tract and are usually injected intramuscularly. Cephalothin does not cross the blood–brain barrier as readily as cephaloridine, but it still enters to a small extent, even when the meninges are inflamed. Cephalexin is absorbed from the gut and so can be given orally.

The metabolism of penicillins by the body is poorly understood but metabolism may occur in the liver. Small amounts of penicillin may be secreted in the bile, but most is excreted in the urine, predominantly by active secretion by the kidney tubules. Probenecid blocks secretion and is used to slow excretion and prolong the duration of action.

The excretion of cephalosporins is rapid, and for cephalothin takes place mainly by secretion, whereas cephaloridine is excreted in the glomerular filtrate. Cephalosporins initiate sensitization less frequently than penicillins although skin rashes occasionally occur, and these drugs rarely cause hypersensitivity responses in individuals who are sensitized to penicillins. Concentrations of cephalexin in the plasma are considerably higher than those of cephalothin or cephaloridine.

3.7 Undesirable effects of the β-lactams

β-Lactams do not interact directly with any intermediary metabolic process in man, but adverse reactions sometimes occur. Acylated proteins (e.g. PBPs) presumably act as haptens. Topical application is limited because of frequent skin rashes. Gastrointestinal discomfort, nausea, vomiting and diarrhea can follow oral administration, perhaps as a result of changes in the gut flora. Hypersensitivity is usually a response to all penicillins, not just to the first form used. The serum-sickness type of reaction is characterized by rashes, urticaria and fever a few hours or days after administration. This response indicates that a more serious anaphylactoid response may occur if a penicillin is used subsequently. This most serious effect follows soon after administration, and its symptoms include cardiovascular collapse, hypotension, weak pulse and shock. In those individuals known to be sensitive to one penicillin, tests for sensitivity to other penicillins may cause anaphylactic reactions. Most people have some antibodies to the drugs as a result of exposure to medicines or residues in foods, milk, or the atmosphere of hospitals. The incidence of hypersensitivity is reduced by the use of highly purified penicillins.

Large intrathecal injections have caused loss of consciousness and convulsions, but this may be due to components other than the peni-cillin in the injection. Renal damage has been caused by doses above 8 g of cephaloridine daily. Less renal damage is reported for cephalothin, and cephalexin appears to be free of this adverse effect.

3.8 Resistance to β-lactams and strategies for treating infections with resistant bacteria

Resistance to β-lactams [39, 40] may be due to hydrolysis of the lactam ring by a β-lactamase enzyme, to the acquisition of high-molecular-weight PBPs with a reduced affinity for the drugs [41], or to the reduced permeability of the outer membrane resulting from changes in the porins [42] of Gram-negative cell walls.

3.8.1 The β-lactamase enzymes

The β-lactamase enzymes [43, 44] were described in the 1940s before the use of penicillin became widespread. Those that hydrolyze penicillins are called penicillinases, and those that hydrolyze cephalosporins are called cephalosporinases. Because these two substrate specificities may not coincide in any one β-lactamase, penicillinase-producing bacteria

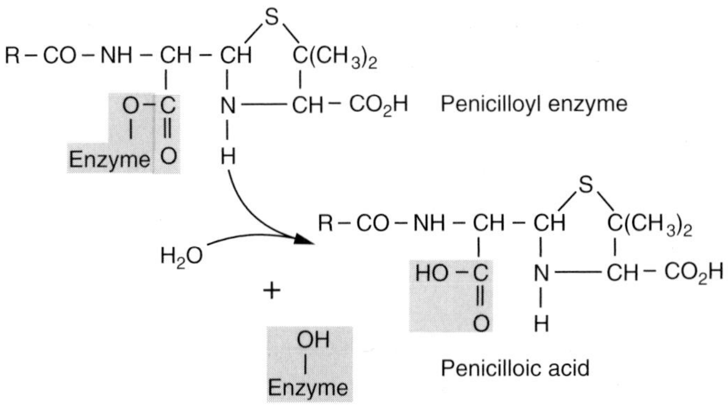

Figure 3.14: Hydrolysis of penicilloyl enzyme complex by β-lactamase.

often remain sensitive to cephalosporins and vice versa. A considerable amount of structural information is now available about these proteins [45]. β-Lactamases occur in both Gram-negative and Gram-positive bacteria exposed to penicillins with increasing frequency [46, 47]. They are penicilloyl serine transferases and react like the transpeptidases to produce penicilloyl–enzymes, but these complexes are readily hydrolyzed to release free enzyme molecules and inactive penicilloic acid (*Figure 3.14*). Mutant enzymes of this type therefore destroy the drugs by hydrolyzing the β-lactam ring. Different classes of β-lactamase have been identified, of which class A enzyme mutants have been the most extensively studied [48]. Class B β-lactamases are zinc enzymes. The three classes of serine-dependent enzymes (A, C and D) are a superfamily that includes the PBPs.

One strategy for dealing with the problem posed by these enzymes is to treat resistant bacteria with a β-lactam drug for which the β-lactamase enzyme has a poor affinity (i.e. to use a cephalosporin for a penicillinase-producing strain and vice versa). A second strategy is the development of semi-synthetic drugs (*Figure 3.15*) that are poor substrates for these enzymes. A wide range of drugs modified in this way is now available, and many of them have a wider range of action or resistance to β-lactamase, or both.

A third option is the use of a combination of a β-lactam drug and an inhibitor of β-lactamase [49, 50], such as clavulanic acid (see Section 3.8.3). Desirable features for the future include greater inhibitory activity against *Pseudomonas* and against MRSA, improved pharmacokinetics that permit doses to be administered only once a day, and even higher degrees of stability to β-lactamases. The phylogenetic relationships between the lactamases are now becoming clear [51, 52].

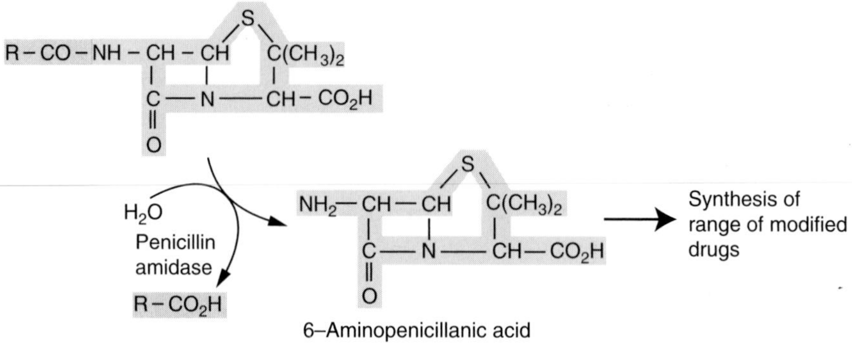

$$R-CO-NH-CH-CH \quad C(CH_3)_2$$
$$C-N-CH-CO_2H$$
$$O$$

H_2O
Penicillin
amidase

$R-CO_2H$

$$NH_2-CH-CH \quad C(CH_3)_2$$
$$C-N-CH-CO_2H$$
$$O$$

Synthesis of range of modified drugs

6–Aminopenicillanic acid

Figure 3.15: Production of the penicillin nucleus by penicillin amidase.

3.8.2 *Semi-synthetic penicillins and cephalosporins*

Semi-synthetic penicillins are prepared by treating penicillins produced by fermentation with a microbial enzyme, penicillin amidase (*Figure 3.15*) which removes the side-chain hydrolytically. A wide range of acyl side-chains have been chemically coupled to the free amino group of the penicillin nucleus. Drugs that have a broader range of antimicrobial activity in some cases, or a reduced sensitivity to penicillinase in others, have been produced in this way. The acyl groups of four semi-synthetic penicillins are illustrated in *Figure 3.16*.

Ampicillin is susceptible to staphylococcal β-lactamase, but is effective against Gram-negative bacteria. Methicillin [53] was the first semi-synthetic pencillin with reduced sensitivity to staphylococcal penicillinase. Cloxacillin and oxacillin are more stable and have better pharmacological properties than methicillin. The penicillinase enzyme has such a low affinity (high K_m) for these molecules that the velocity of hydrolysis is low at therapeutic concentrations. Synergy was reported between ampicillin and cloxacillin in a 1:1 combination, such that 10% or 5% of the concentration of the combination was required to produce the same effect as ampicillin alone for some isolates. Penicillin-resistant pneumococci vary in prevalence worldwide, and are often resistant to non-β-lactam antibiotics. The mechanism of resistance lies in the alteration of PBPs. Resistant strains are commonly carried by children who have recently been hospitalized and recently been exposed to antibiotics.

The development of strains of the common respiratory tract pathogens *H. influenzae* and *Moraxella catarrhalis* that produce β-lactamase has caused increasing resistance to ampicillin and amoxicillin, which are commonly used to treat respiratory tract infections. Therapy at this site is usually empirical, but an understanding of the current resistance profile and its implications for therapy is critical for successful treatment.

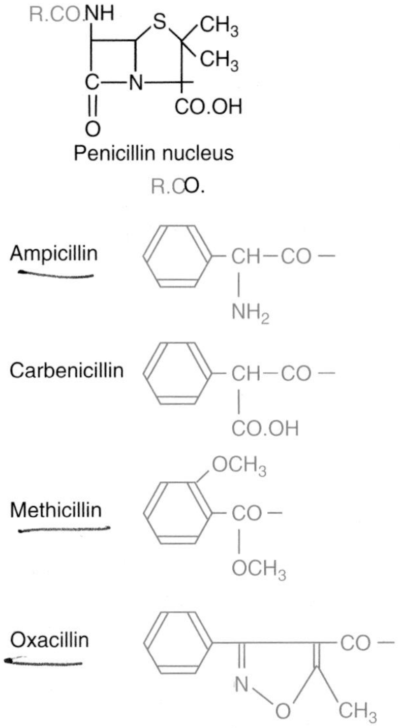

Figure 3.16: Semi-synthetic penicillins.

Piperacillin and related ureido penicillins are effective against Gram-positive and Gram-negative bacteria including *Pseudomonas*, although they are hydrolyzed by Gram-negative penicillinases.

Semi-synthetic cephalosporins are produced in the same way as semi-synthetic penicillins by modifying the acyl group $R_1.CO$- that corresponds to the acyl group of the penicillins (*Figure 3.17*). Cephalosporins also have an acyl group $R_2.CO$-. A class of naturally occurring β-lactams, the cephamycins, that resemble the cephalosporins but have another methoxy group, have an extended useful spectrum as the methoxy substituent confers resistance to the β-lactamases of Gram-negative bacteria.

3.8.3 Inhibitors of β-lactamases

Drugs that inhibit β-lactamases extend the life of drugs that are hydrolyzed by these enzymes. The first indication that some drugs might inhibit the β-lactamases was provided by the semi-synthetic penicillins. (*Table 3.3*). There followed a purposeful search for inhibitors that might not necessarily be antibiotics. The olivanic acid group have not proved useful because they are unstable. Clavulanic acid, oxapenem, sulbactam

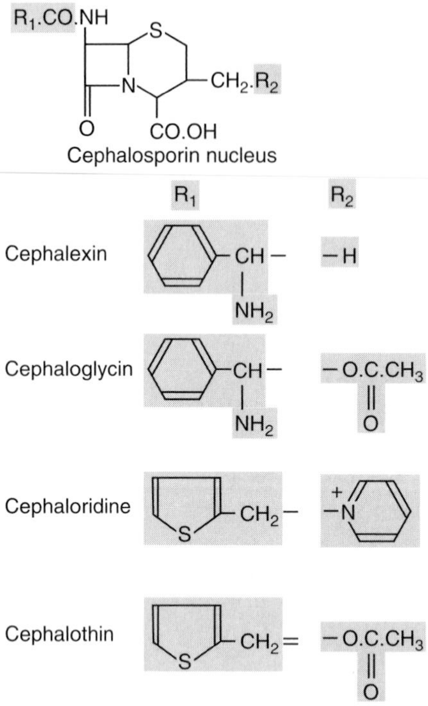

Figure 3.17: Semi-synthetic cephalosporins.

and tazobactam are inhibitors that have been used more successfully (*Figure 3.18*). Although clavulanic acid has no marked antibiotic activity, it reacts with β-lactamase in a similar way to the antibiotics to give a stable acyl-enzyme (*Figure 3.19*). The interaction of clavulanate with the enzyme has been deduced and significant amino acid residues have been identified. A structural study indicates that tazobactam may cross-link Ser70 to Lys73 in the *S. aureus* PC1 β-lactamase [54].

3.8.4 *Combinations of a β-lactam drug and an inhibitor of β-lactamase*

Oral amoxycillin/clavulanic acid (augmentin) is a convenient combination of an inhibitor with the β-lactam drug, but is only one of several available [55].

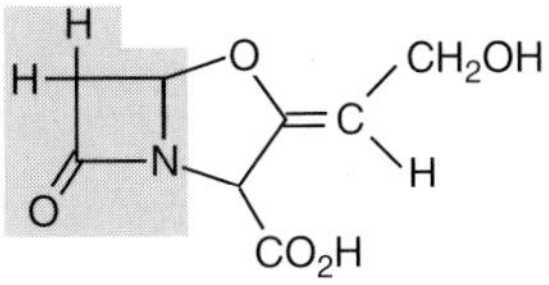

Clavulanic acid

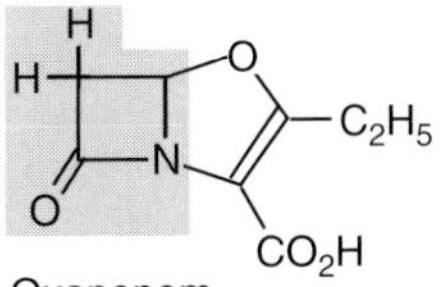

Oxapenem

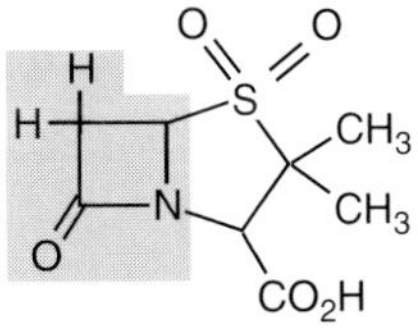

Sulbactam

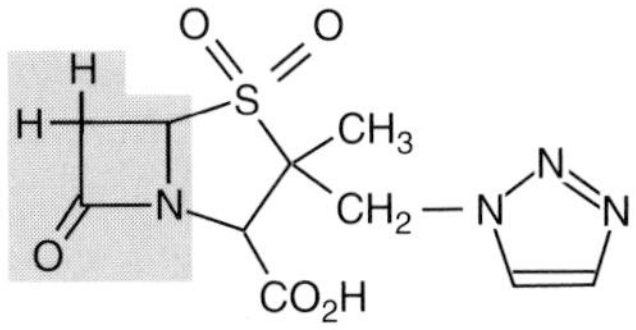

Tazobactam

Figure 3.18: Inhibitors of β-lactamase.

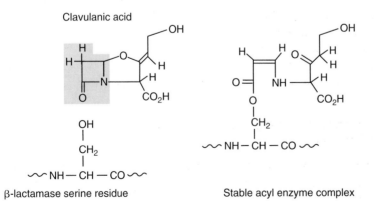

Clavulanic acid

β-lactamase serine residue

Stable acyl enzyme complex

Figure 3.19: Mode of action of clavulanic acid.

3.9 Methicillin-resistant *Staphylococcus aureus* (MRSA)

Approximately 70% of hospital isolates of *S. aureus* are now resistant to penicillins, including methicillin, because they produce potent β-lactamase with wide specificity [56, 57]. The only effective drug available to treat infections with MRSA is vancomycin.

References

1. Tomasz, A. (1980) *Phil. Trans. R. Soc. Lond.*, **B289**, 303–308.
2. Baddiley, J. and Abraham, E.P. (1980) *Penicillin Fifty Years after Fleming*. The Royal Society, London.
3. Spratt, B.G. (1980) *Phil. Trans. R. Soc. Lond.*, **B289**, 273–283.
4. Abraham, E.P. (1981) *Sci. Am.*, **244**, 67–74.
5. Ghuysen, J.M. (1990) *Biotechnol. Appl. Biochem.*, **12**, 468–472.
6. Ward, J.B. (1990) in *Comprehensive Medicinal Chemistry*, Vol. 2 (C. Hansch P.G. Sammes and J.B. Taylor, eds). Pergamon, Oxford, pp. 553–607.
7. Ghuysen, J.-M. and Hakenbeck, R. (1994) in *Bacterial Cell Wall. New Comprehensive Biochemistry*, Vol. 27 (A. Neuberger and L.L.M. van Deenen, eds). Elsevier, Amsterdam, pp. 1–581.
8. Faraci, W.S. (1992) in *Emerging Targets in Antibacterial and Antifungal Chemotherapy*, (J.A. Sutcliffe and N.H. Georgopapadakou, eds). Chapman & Hall, New York, pp. 176–204.
9. Kahan, F.M., Kahan, J.S., Cassidy, P.J. and Kropp, H. (1974) *Ann. N.Y. Acad. Sci.*, **235**, 364.
10. Bergan, T. (1990) *Chemotherapy*, **36** (Suppl.1), 10–18.
11. Walsh, C.T. (1989) *J. Biol. Chem.*, **264**, 2393–2396.
12. Lambert, M.P. and Neuhaus, F.C. (1972) *J. Bacteriol.*, **109**, 1156–1161.
13. Lambert, M.P. and Neuhaus, F.C. (1972) *J. Bacteriol.*, **110**, 978–987.
14. Mendoza, M.C., Teran, F.J., Mendez, F.J. and Hardisson, C. (1988) *Microbiologica*, **11**, 289–297.
15. Etienne, J., Gerbaud, G., Fleurette, J. and Courvalin, P. (1991) *FEMS Microbiol. Lett.*, **68**, 119–122.
16. Stone, K.J. and Strominger, J.L. (1971) *Proc. Natl Acad. Sci. USA*, **68**, 3223–3227.
17. Storm, D.R. (1974) *Ann. N.Y. Acad. Sci.*, **235**, 387–398.
18. Waltho, J.P. and Williams, D.H. (1991) *Ciba Found. Symp.*, **158**, 73–86.
19. Nagarajan, R. (1991) *Antimicrob. Agents Chemother.*, **35**, 605–609.
20. Bochud-Gabellon, I. and Regemey, C. (1992) *Dermatologia*, **176**, 29–38.
21. Hakenbeck, R. (1994) in *Bacterial Cell Wall* (J.-M. Ghuysen and R. Hackenbeck, eds) Elsevier, New York pp. 535–545.
22. Handwerger, S., Pucci, M.J., Volk, K.J., Liu, J. and Lee, M.S. (1992) *J. Bacteriol.*, **174**, 5982–5984.
23. Courvalin, P. (1990) *Antimicrob. Agents Chemother.*, **34**, 2291–2296.
24. Dutka-Malen, S., Molinas, C., Arthur, M. and Courvalin, P. (1992) *Gene*, **112**, 53–58.

25. Bugg, T.D., Dutka-Malen, S., Arthur, M., Courvalin, P. and Walsh, C.T. (1991) *Biochemistry*, **30**, 2017–2021.
26. Wright, G.D., Molinas, C., Arthur, M., Courvalin, P. and Walsh, C.T. (1992) *Antimicrob. Agents Chemother.*, **36**, 1514–1518.
27. Newall, C.E. and Hallam, P.D. (1990) in *Comprehensive Medicinal Chemistry*, Vol. 2 (C. Hansch P.G. Sammes J.B. Taylor, eds). Pergamon, Oxford, p. 609–653.
28. Brown, A.G., Pearson, M.J. and Southgate, R. (1990) in *Comprehensive Medicinal Chemistry*, Vol. 2 (C. Hansch, P.G. Sammes and J.B. Taylor, eds). Pergamon, Oxford, pp. 655–702.
29. Kahan, J.S., Kahan, F.M., Goegelman, R. *et al.* (1979) *J. Antibiot.*, **32**, 1–12.
30. Boyd, D.B. (1977) *Proc. Natl Acad. Sci USA*, **74**, 5239–5243.
31. Waxman, D.J. (1980) *Phil. Trans. R. Soc. Lond.*, **B289**, 257–271.
32. Spratt, B.G. (1983) *J. Gen. Microbiol.*, **129**, 1247–1268.
33. Spratt, B.G. and Crombie, K.D. (1988) *Rev. Infect. Dis.*, **10**, 699–711.
34. Tamura, T. (1980). *Proc. Natl Acad. Sci. USA*, **77**, 4499–4503.
35. Tuoman, E. and Schwartz, J. (1987) *J. Bacteriol.*, **169**, 4912–4915.
36. Tomasz, A. and Waks, S. (1975) *Proc. Natl Acad. Sci. USA*, **72**, 4162–4166.
37. Tomasz, A. (1980) *Phil. Trans. R. Soc. Lond.*, **B289**, 303–308.
38. Holtje, J.V. and Tomasz, A. (1975) *Proc. Natl Acad. Sci. USA*, **72**, 1690–1694.
39. Ghuysen, J.M. (1988) *Rev. Infect. Dis.*, **10**, 726–732.
40. Malouin, F. and Bryan, L.E. (1986) *Antimicrob. Agents Chemother.*, **30**, 1–5.
41. Joris, B., Hardt, K. and Ghuysen, J.M. (1994) in *Bacterial Cell Wall* (J.-M. Ghuysen and R. Hakenbeck, eds). Elsevier, New York, pp. 505–515.
42. Chou, J.H., Greenberg, J.T. and Demple, B. (1993) *J. Bacteriol.*, **175**, 1026–1031.
43. Ghuysen, J.M. (1990) *Biotechnol. Appl. Biochem.*, **12**, 468–472.
44. Ghuysen, J.-M. and Dive, G. (1994) in *Bacterial Cell Wall* (J.-M. Ghuysen and R. Hakenbeck, eds.). Elsevier, New York, pp. 103–129
45. Knox, J.R. (1993) in *Recent Advances in the Chemistry of Anti-infective Agents* (P.H. Bentley and R. Ponsford, eds). Royal Society of Chemistry, London.
46. Sanders, C.C. (1990) *Clin. Infect. Dis.*, **14**, 1089–1099.
47. Sanders, C.C. and Sanders, W.E (1992) *Clin. Infect. Dis.*, **15**, 824–839.
48. Matagne, A. and Frere, J.-M. (1995) *Biochim. Biophys. Acta*, **1246**, 109–127.
49. Rolinson, G.N. (1991) *Rev. Infect. Dis.*, **13** (Suppl. 9), S727–S732.
50. Moellering, R.C. (1991) *Rev. Infect. Dis.*, **13** (Suppl. 9), S723–S726.
51. Ogawara, H. (1993) *Mol. Phylogenet. Evol.*, **2**, 97–111.
52. Spratt, B.G. (1994) in *Bacterial Cell Wall* (J.-M. Ghuysen and R. Hakenbeck, eds). Elsevier, New York. p. 517.
53. Spratt, B.G. (1977) *J. Antimicrob. Chemother.*, **3** (Suppl. B), 13–19.
54. Denny, B.J., Toomer, C.A. and Lambert, P.A. (1994) *Microbios*, **78**, 245–257.
55. Citron, D.M., Goldstein, E.J.C., Kenner, M.A., Burnham, L.B. and Inderlied, C.B. (1995) *Clin. Infect. Dis.*, **20** (Suppl. 2), S356–S360.

56. Cafferkey, M.T (1992) *Methicillin-resistant* Staphylococcus aureus. Dekker, New York.

57. Mulligan, M.E., Murray-Leisure, K.A., Ribner, B.S., Standiford, H.C., John, J.F., Korvick, J.A., Kaufman, C.A. and Yu, V.L. (1993) *Am. J. Med.*, **94**, 313–328.

Chapter 4

Synthesis of nucleic acids as a target for antimicrobial drugs

4.1 Introduction to inhibition of nucleic acid synthesis

The synthesis of deoxyribonucleic acid (DNA) and ribonucleic acid (RNA) is essential for the growth and maintenance of all living cells. The formation of nucleic acids may be prevented at several stages (*Figure 4.1*). Analogs (antimetabolites) of essential metabolites may

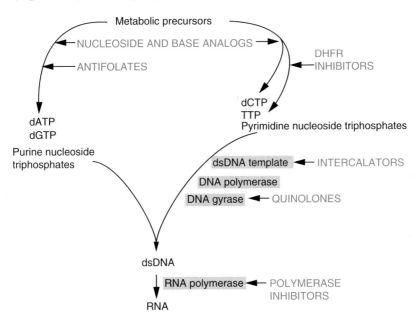

Figure 4.1: Points at which drugs may inhibit nucleic acid synthesis.

61

inhibit the synthesis of the nucleotide bases, or the nucleosides or nucleotides, which represent the substrates from which the macro-molecules are made. Drugs that bind to the DNA double helix may block its template activity in the replication of new DNA molecules, or in the transcription of DNA to form RNA. Nucleic acid synthesis can also be blocked by drugs that bind to and interfere with the action of DNA-dependent RNA polymerase or DNA-dependent DNA poly-merase. Alternatively, DNA synthesis can be inhibited by drugs that disable one of the several accessory proteins and enzymes involved in DNA replication.

In each case there is a requirement for selectivity. In order to be useful clinically, a drug must inhibit the process of nucleic acid synthesis in an infective micro-organism at a concentration that, under normal circumstances, has little or no effect on the human tissues.

4.2 Drugs effective against the early stages of nucleic acid synthesis

4.2.1 Antimetabolites have few applications in antimicrobial therapy

The prevention of nucleotide synthesis by the antifolate drugs such as methotrexate, trimethoprim and sulfonamides (Chapter 2) has already been described. One of the actions of these drugs is to block the forma-tion of deoxythymidylate monophosphate (dTMR), a precursor of DNA. Azaserine and diazo-oxo-norleucine (DON) are antibiotics that inhibit the transfer of the amide nitrogen of glutamine during the synthesis of purine bases (*Figure 4.2*). They have been used in tumor therapy and as antiviral agents (as have analogs of nucleosides and their bases), but are not useful in antimicrobial chemotherapy because their lack of selec-tivity causes them to be toxic to the host tissues.

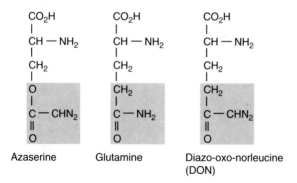

Azaserine   Glutamine   Diazo-oxo-norleucine
                                       (DON)

Figure 4.2: Antimetabolites effective in nucleotide synthesis.

4.2.2 Intercalators and their effects on DNA and RNA synthesis

Acridine drugs were widely used as topical wound disinfectants during the First World War, and are commonly used as the active ingredients in the yellow ointments and creams supplied in first-aid kits for use on small burns and abrasions. Polycyclic aromatic compounds of this type (*Figure 4.3*) have been employed as vital stains for eukaryotic tissues because, when they bind to DNA, there are significant changes in their spectral properties. These dyes are flat molecules that intercalate between the base pairs of DNA. They destroy the typical X-ray diffraction pattern of double-helical DNA fibers because they disrupt the regularity of the structure. Intercalating drugs also increase the viscosity of DNA solutions by increasing the effective length of the DNA helices, and decrease their buoyant density because the molecular weights of the acridines are less than those of the base pairs that they mimic. In addition, by holding the strands of the double helix together, through interaction with the bases of both strands, intercalating drugs increase the thermal stability of DNA.

Acridines were also the mutagens used to produce the 'frameshift' mutants of bacteriophage (virus) T4 that provided evidence for the triplet nature of the genetic code. An insertion mutant may be rendered similar to the wild type by a compensating deletion or two further insertions near the mutation. Such compounds do not often cause mutagenesis in most bacteria because mutations only appear to occur when crossing-over of chromosomes takes place, which is a rare event in bacteria, except during the conjugation of Gram-negative bacteria. Acridine treatment causes the loss of plasmids (extrachromosomal

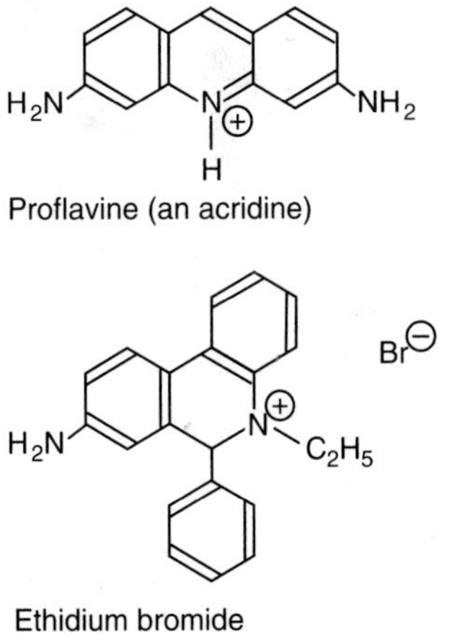

Proflavine (an acridine)

Ethidium bromide

Figure 4.3: Intercalating drugs.

DNA) in bacteria, the loss of proteins specified by mitochondrial DNA, and the loss of the kinetoplast of trypanosomes, all of which involve failure to replicate small closed circles of DNA.

Ethidium bromide (*Figure 4.3*) is used as a reagent for the microassay of DNA, and for visualizing nucleic acids in electrophoretic separations of nucleic acid fragments in molecular biology laboratories. This is because of the enhancement of its fluorescence on binding to double-stranded DNA. Ethidium bromide has been used against trypanosomes in veterinary practice.

Treatment of bacterial cells with intercalating drugs inhibits both DNA polymerase and RNA polymerase, although one or the other may be inhibited preferentially.

4.2.3 Undesirable effects of drugs acting at early stages of synthesis

Intercalating drugs are of many types and some of them (e.g. acriflavine and proflavine) are used in general antiseptic creams for topical application. Many intercalators are too toxic for systemic use, but an acridine drug (mepacrine) has been used against the plasmodium parasite of malaria, and ethidium bromide is useful against trypanosomes, both organisms being protozoa.

4.2.4 Actinomycin D (dactinomycin) and RNA synthesis

This antibiotic (*Figure 4.4*) binds to DNA, particularly at guanosine–cytosine (G+C) base pairs, via the amino and keto groups at the side of the ring systems. One model for the binding places the two hydrophilic rings containing amino acids in an external groove on the DNA helix. The most important biological function of actinomycin D is the selective inhibition of RNA synthesis. The degree of inhibition depends on the ratio of antibiotic to DNA and not on the concentration of enzyme or of nucleotides, indicating that the effect is exerted directly on the DNA template. Synthetic double-stranded polydeoxy(AT) contains no G or C bases. RNA synthesis by RNA polymerase directed by such a template, which does not bind actinomycin D, is not sensitive to the drug. The inhibition of RNA polymerase action takes place at low concentrations of actinomycin D that do not increase the thermal stability of DNA, indicating that intercalation is not significant in the process. At higher actinomycin D concentrations, the thermal stability of the nucleic acid is increased, and the action of DNA polymerase is inhibited. It has been postulated that DNA polymerase activity involves an interaction between the enzyme and the double helix at the wide groove, while RNA polymerase action takes place in the narrow groove. Actinomycin D is toxic to eukaryotic cells and is not used for anti-

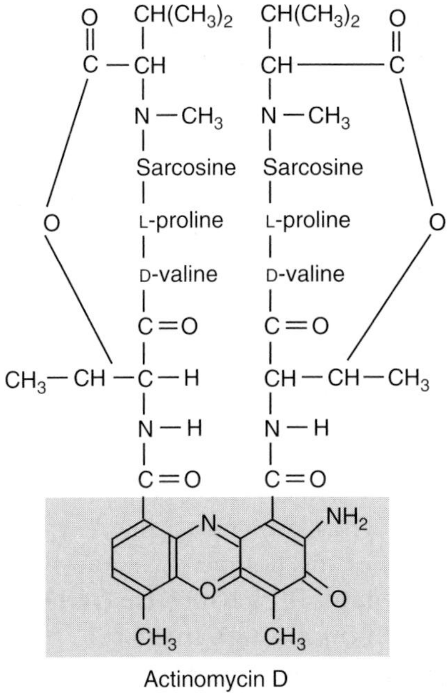

Actinomycin D

Figure 4.4: Actinomycin D.

microbial chemotherapy. It has been used to treat Wilms tumor of the kidney in children. The drug is used predominantly for experimental purposes (e.g. to block the synthesis of further new RNA after the incorporation of radioactive precursors during pulse labeling experiments). Actinomycin D is of no use in the chemotherapy of infections.

4.3 The role of topoisomerases in DNA synthesis

Cells require enzymes that assist with the writhing and twisting of the huge unwieldy molecules of DNA during both replication of DNA and the expression of the genetic message as mRNA. Because the bacterial chromosome consists of closed circles of double helix, topological problems occur in the replication of daughter molecules. The chromosome is 1000-fold longer than the cell. In the DNA double-helical configuration of Watson and Crick (relaxed conformation), one strand passes around the other once every 10 bp. If the circle has 4×10^6 bp in the *E. coli* chromosome, the strands are intertwined 4×10^5 times (the linkage number) if the DNA is in the relaxed conformation. These turns must be unwound if the chromosome is to replicate. If the cell divides once every 40 min, the DNA unwinding rate must be 10 000 turns min^{-1}. The DNA conformation is often not relaxed, but supercoiled. Supercoiling may be positive (i.e. coiled in the same sense as the turns of the double helix). This makes the DNA more compact and prevents replication and transcription. Negative supercoiling facilitates separation of the two strands and is required for the action of both DNA and RNA polymerases. Enzymes have been discovered that can nick and reseal the strands of DNA and change its topology. They exist in all cells, where they facilitate DNA replication and gene expression. These enzymes are called 'topoisomerases' [1–3]. During the transcription of DNA into mRNA, positive supercoils accumulate ahead of the RNA polymerase as it traverses double-stranded DNA, and failure of gyrase action to release this would stop transcription.

4.3.1 *Topoisomerases, particularly DNA gyrase*

Topoisomerases (*Table 4.1*) cut one or both strands of dsDNA, and either facilitate unwinding or drive supercoiling. They are of two main types. Type I enzymes transiently nick one strand of the double helix and pass the other strand through the gap, whereas type II enzymes cleave both strands and pass another double-helical section through the gap.

The type I topoisomerase of *E. coli* is a monomeric 97-kDa zinc metalloprotein that has a 24% homology over the 300 amino acid residues, including the active site tyrosine residue, with topoisomerase III. These are both type I enzymes that can remove negative, but not positive superhelicity without the hydrolysis of ATP.

Topoisomerase II or DNA gyrase [4] can introduce or remove supercoiling, cause circles of DNA to be concatenated or decatenated, as well as knotting or unknotting DNA (*Figure 4.5*), can separate the two daughter chromosomes before cell division, and relieve the positive supercoiling that is caused by the progress of the replicating fork (*Figure 4.6*). The structure and function of DNA gyrase has been extensively reviewed [4]. It is the enzyme that can introduce negative supercoiling into DNA *in vivo* with the hydrolysis of ATP. The two peptides of this $\alpha_2\beta_2$ tetrameric enzyme (coded by genes *gyrA* and *gyrB)* are of

Table 4.1: Topoisomerase enzymes of *E. coli*

Number	Type	Gene	M_r (kDa)	Amino acids	Enzyme activities	ATP requirement
I	I	*topA*	110		Removes negative supercoil	–
II	II	*gyrA* *gyrB*	97 90	875 804	Holoenzyme both removes positive super-coiling and introduces negative supercoil	– +
III	I	*topB*			Removes negative supercoil	–
IV	II	*parC* *parE*				

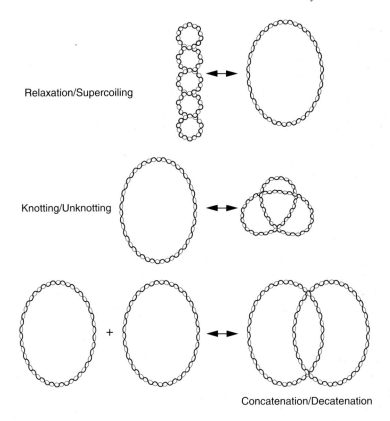

Figure 4.5: The reaction catalyzed by DNA gyrase.

molecular mass (M_r) 97 kDa (875 amino acids) and 90 kDa (804 amino acids), respectively. DNA gyrase, which is not present in eukaryotic cells, is the site of action of a series of drugs, including both natural and synthetic compounds. Mutations in genes for the A and B subunits code for resistance to nalidixic acid and coumermycin, respectively. Gyrase cuts both strands and, in so doing, changes the linkage number by two. The strands are cut with a 4 nucleotide stagger, and each 5' terminus is transiently covalently linked to Tyr122 of the A subunits. The quinolone drugs bind to the exposed single-strand termini of the complex [5, 6]. The drugs do not bind to either subunit of the purified enzyme without DNA, but they do bind weakly to DNA without the enzyme.

4.3.2 The quinolones

Nalidixic acid is a synthetic quinolone compound (*Figure 4.7*) and was described as a new class of chemotherapeutic drug in 1962. The quinolones inhibit the replication of DNA without immediately affecting

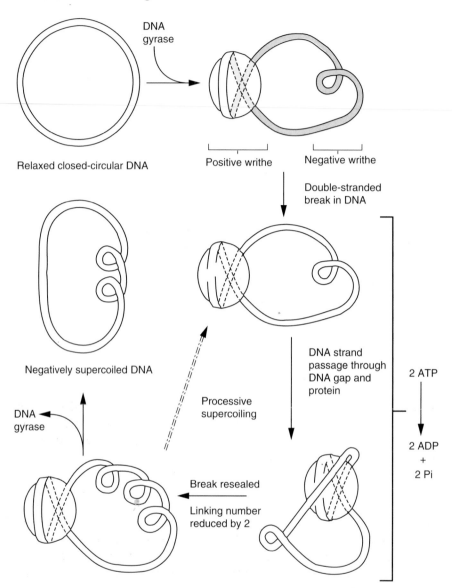

Figure 4.6: The mechanism of action of DNA gyrase. Reproduced from Reece, R.J., Maxwell, A., *Gyrase: Structure and Function*, 354, in *Critical Reviews in Biochemistry, and Molecular Biology,* Vol. 26, Iss. 3, 4 Fasman, G. D., Ed., CRC Press, Boca Raton, Florida, © 1991. With permission.

RNA or protein synthesis in sensitive bacteria. Nalidixic acid causes disintegration of DNA and filamentation of bacterial cells, but its toxicity in animals is limited to inhibition of mitochondrial DNA replication. The mechanism of action of quinolones was tested with plasmids, which

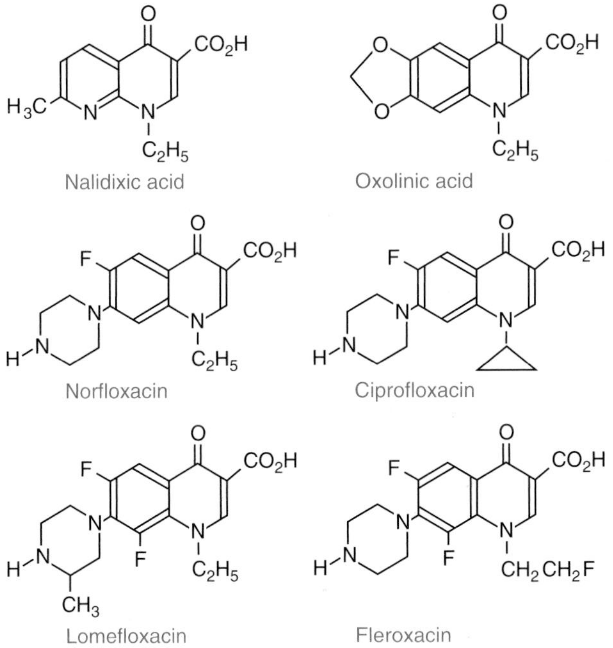

Figure 4.7: Structures of the quinolones.

are converted from a circular to a linear conformation. The drug binds to the larger subunit of the tetrameric topoisomerase DNA gyrase, and mutations to resistance (nal[R]) in the *nalA* (*gyrA*) gene of *E. coli* result in an A protein that does not bind to the inhibitor.

Nalidixic acid is excreted in the urine, together with inactive meta-bolites, and has now been superseded because it does not achieve bactericidal concentrations in any fluid except urine. It is bactericidal to most of the Gram-negative bacteria that cause the common urinary tract infections, with the exception of *Pseudomonas* strains.

Oxolinic acid (*Figure 4.7*) and cinoxacin are members of the first gen-eration of quinolones, and are both more inhibitory than nalidixate and effective against a wider range of bacteria. Norfloxacin and ciprofloxacin are members of the second generation of 6-fluoroquinolones [7] that are more effective against a wider range of bacteria. Lome-floxacin and fleroxacin are examples of more recent difluoro com-pounds. These newer quinolones are active against a wide range of Gram-negative aer-obes and moderately effective against Gram-positive aerobes.

4.3.3 Administration of nalidixic acid and the quinolones

Nalidixic acid is absorbed from the gastrointestinal tract, but a large proportion of the drug is bound to plasma proteins and penetration into

Table 4.2: Effects of quinolones on cells and peak serum concentration

	Minimum inhibitory concentration (μg ml^{-1})	Maximal bactericidal concentration (μg ml^{-1})	Peak serum concentration (μg ml^{-1})
Nalidixic acid	3.0	90.0	20–50
Ofloxacin	0.03	0.9	11
Norfloxacin	0.04	1.5	1.5
Ciprofloxacin	0.004	0.15	1.9–2.9

the tissues is poor. The plasma half-life is approximately 1.5 h, and the major route of excretion of both the drug and its metabolites is via the kidney. The metabolites of nalidixic acid are bacteriologically inactive, but sufficient levels of unmetabolized compound are excreted into the urine to make nalidixic acid useful in the treatment of urinary tract infections. The newer quinolones have improved selectivity (*Table 4.2* and *4.3*) and are better absorbed and less rapidly cleared than nalidixic acid so dosing need only take place at 12- to 24-h intervals. Urinogenital, gastrointestinal, respiratory, skin and bone infections may all be treated. The pharmacokinetics of these compounds have been extensively reviewed [8].

4.3.4 Undesirable effects of the quinolones

The toxic effects of quinolones include nausea and vomiting, skin rashes and photosensitivity reactions, CNS effects such as dizziness and visual disorders, and convulsions. Mitochondrial DNA replication is inhibited at concentrations that do not affect the nuclear DNA. Nalidixic acid can displace warfarin from protein binding sites and can prolong the clotting time and cause hemorrhages, if these drugs are administered together.

Temefloxacin, which was promising as a treatment for respiratory infections caused by Gram-positive bacteria, was withdrawn because of a rare antibody-dependent reaction.

4.3.5 Resistance to the quinolones

There are two distinct mechanisms of resistance to quinolones: mutations in DNA gyrase and changes in cell permeability to the drugs. There is no example described in which quinolones are modified or destroyed. The target-mediated resistance most commonly involves a mutation in the GyrA protein subunit leading to reduced binding affinity for the quinolones. Target mutations are less commonly due to mutation in the GyrB protein, and of 40 *E. coli* mutants only four were in the GyrB subunit [9]. Gyrase B subunit mutations may enhance the effect of mutations in GyrA and may affect the interactions between the two

Table 4.3: Effects of quinolones on two topoisomerases

	IC$_{50}$ (µg ml^{-1})		Selectivity ratio
	E. coli gyrase	Calf thymus topo II	
Nalidixic acid	23	385	17
Lomefloxacin	0.78	280	359
Ciprofloxacin	0.13	155	1192
Ofloxacin	0.76	1870	2461

subunits of the enzyme. In fluoroquinolone-resistant *S. typhimurium*, the introduction of plasmids bearing either the *E. coli gyrA* or *gyrB* gene conferred some susceptibility, but both genes were able to restore wild-type sensitivity to quinolones [10]. Resistance to nalidixic acid develops readily but is apparently not transferable from one organism to another. The MICs for the drugs are increased by 2- to 128-fold by mutations that cluster between residues Ala67 and Gln106 of the 875 residue protein. Residue Ser83 mutations confer the greatest resistance. In *M. tuberculosis* strains that had a ciprofloxacin MIC of more than 2 µg ml^{-1} there were mutations of *gyrA* analogous to those described in other fluoroquinoline-resistant bacteria [11]. *P. aeruginosa* has a gyrase A protein that is 67% identical with that of *E. coli*, and in ciprofloxacin-resistant mutants Asp87 is replaced by Asn or Thr83 by Ile [12].

The second method of resistance to quinolones is by mutational changes in the properties of the outer cell membrane that render it less permeable to the drugs [13]. Penetration of the outer wall of *E. coli* is via the water-filled OmpF and OmpC porin channels, but permeability is less dependent on the hydrophobicity of the quinolones than on their ionization. OmpC mutants show no increase in drug sensitivity. Mutations *norB*, *cfxB* and *marA* are distinct from these porins and may cause decreased susceptibility to tetracyclines and chloramphenicol as well as to quinolones. In *P. aeruginosa* the expression of OprK in the outer membrane is associated with resistance to quinolones, tetracyclines and chloramphenicol [14]. Mutants deficient in this gene, or two others that form the proteins of a multiple drug efflux system *(mexAB)*, show enhanced sensitivity to all these drugs.

4.3.6 Coumermycin and novobiocin

Novobiocin and coumermycin A (*Figure 4.8*) are effective against the GyrB protein subunit. They inhibit the binding of ATP and the DNA-dependent ATPase activity of the enzyme. Novobiocin (10 nM) raises the K$_m$ for ATP by 1000-fold, and consequently the ATPase activity of DNA gyrase is reduced.

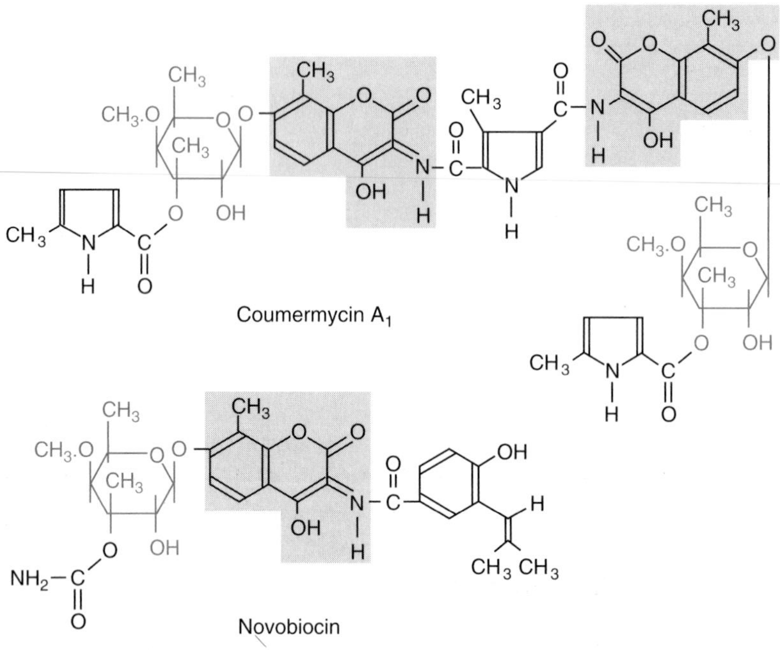

Figure 4.8: Structures of coumermycin and novobiocin.

4.4   Drugs that cause DNA cleavage

Some drugs, or their metabolic products, cleave dsDNA rather than affecting its synthesis. The details of the mechanism of action are rarely clearly established.

4.4.1 Nitroimidazoles

Metronidazole (Flagyl) (*Figure 4.9*) has been used to treat amebic dysentery and infections caused by *Trichomonas vaginalis*. It is also valuable for treating infections caused by strictly anaerobic bacteria (clostridia, fusobacteria and *Bacteroides* species) and for prophylaxis against such infections during abdominal surgery. In the absence of oxygen the drug is reductively converted to a radical ion of the nitro group. This highly reactive species cleaves DNA molecules, particularly those with a high thymine content.

4.4.2 Nitrofurans

Nitrofurantoin (*Figure 4.9*) is extensively and rapidly excreted in the urine, and is effective against most urinary infections, particularly those

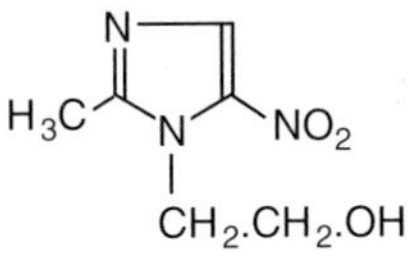

Figure 4.9: Structures of metronidazole and nitrofurantoin.

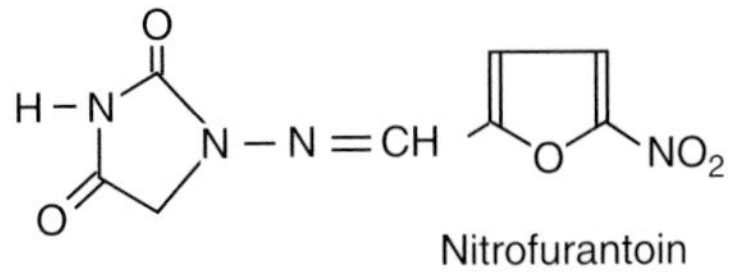

caused by *P. aeruginosa* and *Proteus* species. The mechanism of action of this drug is assumed to resemble that of metronidazole.

4.5 Bacterial RNA synthesis

4.5.1 Bacterial RNA polymerase

The bacterial RNA polymerase (RNA pol) is a large enzyme, of molecular weight almost 500 000, and composed of polypeptide chains of four types, two α, one β, and one β' in the core enzyme molecule. One small subunit, the σ (sigma) factor, binds transiently to the core enzyme to assist binding at promoter sites in DNA, so that RNA synthesis may begin.

4.5.2 Rifamycins and their derivatives

Rifampicin is a semi-synthetic derivative of the complex rifamycin antibiotic nucleus (*Figure 4.10*). Rifapentine is more lipophilic, more slowly cleared from tissues, and is effective against tuberculosis with a weekly dosing regime.

4.5.3 Rifampicin

Rifampicin inhibits the growth of most Gram-positive strains and some Gram-negative strains such as *E. coli*, *Pseudomonas*, *Proteus* and *Klebsiella*. It is the drug of choice for the prophylaxis of meningococcal infections in individuals at risk because they have been in contact with affected persons, and it is also used in the treatment of pulmonary tuberculosis. The drug has no affinity for nucleic acids but binds avidly to DNA-dependent RNA polymerase [15, 16]. The enzyme from *E. coli* is 50% inhibited by 2×10^{-8} M rifampicin (i.e. 10% of the concentration of actinomycin D required to achieve the same effect with *in vitro* RNA

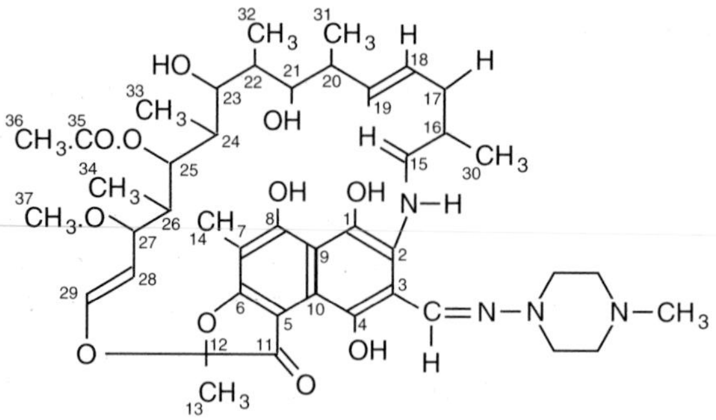

Figure 4.10: Structure of rifampicin.

synthesis). DNA polymerase I is not inhibited at all by 10 000 times this concentration of the drug. RNA polymerase is completely inhibited when one molecule of rifampicin binds per enzyme molecule.

Rifampicin inhibits the transition from state I to state II in RNA polymerase bound at the promoter, and it blocks the synthesis of the first dinucleotide of the new RNA molecule.

4.5.4 Administration of rifampicin

Rifampicin is well absorbed when given orally, and as it is excreted slowly, prolonged therapeutically effective plasma concentrations are achieved. Rifampicin penetrates all body compartments and may cause sweat, tears and urine to be colored pink. The major metabolite of rifampicin, formed in the liver, is as active as the parent compound. Excretion is mainly by secretion into the bile, and as some reabsorption occurs, the enterohepatic circulation may contribute to the long half-life of the drug in the body. Rifampicin accumulates in the body during liver failure or biliary-duct obstruction, but some urinary excretion can take place when plasma concentrations are high.

Rifampicin is used mainly for the treatment of tuberculosis and leprosy. All populations of bacteria contain some individuals which are resistant to the action of rifampicin, and therefore it must always be used in combination with another drug to which the bacteria are sensitive.

4.5.5 Undesirable effects of rifampicin

Adverse reactions to rifampicin are relatively rare, and effects on the liver are frequently reversible if use of the drug is discontinued. The side effects most frequently encountered are rashes, diarrhea, nausea, vomiting and lethargy following large doses of the drug.

4.5.6 *Resistance to rifampicin*

The synthesis of RNA *in vitro* by the DNA-dependent RNA polymerase from rifampicin-resistant strains of bacteria is itself resistant to rifampicin. A β-subunit from a resistant strain is sufficient to confer rifampicin resistance on the reconstituted enzyme, irrespective of the source of the other subunits. Therefore rifampicin binds to the β-polypeptide in order to inhibit the enzyme. Sigma factor and rifampicin both bind independently to the β-protein.

Rifampicin may be drawn in a conformation that resembles two adenosine nucleotides in RNA. The ring system from C_9 to C_{12} corresponds to the base of the first nucleotide. The drug hydroxyls at C_{21} and C_{23} correspond to the hydroxyls of ribose at C_2 and C_3 of the second nucleotide [17]. These hydroxyls in both the drug and the nucleotide form hydrogen bonds with Asp516 of the β-subunit of RNA polymerase. A mutant in which this residue is replaced by the hydrophobic amino acid valine is resistant to rifampicin.

Mycobacteria develop resistance to rifampicin by a rapid one-step process that is a mutation of RNA polymerase, and therefore the drug should be used in combination with others. Most of the mutations to resistance lie in a 50-amino-acid sequence of the 1342-residue subunit. This region of RNA polymerase is part of the rifampicin binding site, as bound rifampicin protects it from proteolysis. Resistant strains have appeared within as short a period as 2 days after the start of treatment. Of 66 resistant *M. tuberculosis* strains, 64 showed mutations among eight highly conserved amino acid residues in a region of the β-subunit that is only 23 residues long [18]. Using this information together with modern genetic techniques of DNA amplification, probing and cloning, strategies to determine which mutations will cause resistance have been developed [19].

References

1. Schmid, M.B. and Sawitzke, J.A. (1993) *BioEssays*, **15**, 445–449.
2. Roca, J. (1995) *Trends Biochem. Sci.*, **20**, 156–160.
3. Luttinger, A. (1995) *Mol. Microbiol.*, **15**, 601–606.
4. Reece, R.J. and Maxwell, A. (1991) *Crit. Rev. Biochem. Mol. Biol.*, **26**, 335.
5. Shen, L.L., Kohlbrenner, W.E., Weigl, D. and Baranowski, J. (1989) *J. Biol. Chem.*, **264**, 2973–2994.
6. Shen, L.L., Mitscher, L.A., Sharma, P.D., O'Donnell, T.J., Chu, D.T.W., Cooper, C.S., Rosen, T. and Pernet, A.G. (1989) *Biochemistry*, **28**, 3886–3894.
7. Piddock, L.V.J. (1993) *ASM News*, **59**, 603–608.
8. Sorgel, F. and Kinzig, M. (1993) *Am. J. Med.*, **94**, 3A45S–3A55S.
9. Herrera, G., Aleixandra, V., Urios, A. and Blanco, M. (1993) *FEMS Microbiol. Lett.*, **106**, 187–191.

10. Heisig, P. (1993) *J. Antimicrob. Chemother.*, **32**, 367–377.
11. Takiff, H.E., Salazar, L., Guerrero, C., Philipp, W., Huang, W.M., Kreiswirth, B., Cole, S.T., Jacobs, W.R. and Telenti, A. (1994) *Antimicrob. Agents Chemother.*, **38**, 773–780.
12. Kureishi, A., Diver, J.M., Beckthold, B., Schollaardt, T. and Bryan, L.E. (1994) *Antimicrob. Agents Chemother.*, **38**, 1944–1952.
13. Dechene, M., Leying, H. and Cullmann, W. (1990) *Chemotherapy*, **36**, 13–23.
14. Poole, K., Krebes, K., McNally, C. and Neshat, S. (1993) *J. Bacteriol.*, **175**, 7363–7372.
15. Sensi, P. and Lancini, G. (1990) in *Comprehensive Medicinal Chemistry*, Vol. 2 (C. Hansch, P.G. Sammes, and J.B. Taylor, eds). Pergamon, Oxford, pp. 793–811.
16. Howe-Grant, M. (1993) *Chemotherapeutics and Disease Control*. Wiley, New York, pp. 140–174.
17. Lal, R. and Lal, S. (1994) *BioEssays*, **16**, 211–216.
18. Telenti, A., Imboden, P., Marchesi, F., Lowrie, D., Cole, S., Colston, M.J., Matter, L., Schopfer, K. and Bodmer, T. (1993) *Lancet*, **341**, 647–650.
19. Miller, L.P., Crawford, J.T. and Shinnick, T.M. (1994) *Antimicrob. Agents Chemother.*, **38**, 805–811.

Chapter 5

Protein synthesis as a target for antibiotics

5.1 Inhibition of protein synthesis in micro-organisms

Inhibitors of RNA synthesis cause a secondary cessation of protein synthesis, but compounds that directly inhibit the assembly of polypeptides have a rapid and specific effect on the accumulation of new protein in growing cells. The addition of an antibiotic such as chloramphenicol to rapidly dividing bacteria stops protein synthesis within a few minutes, while RNA and DNA synthesis continue unaltered for some time (*Figure 5.1*).

Most inhibitors of protein synthesis are bacteriostatic, preventing the growth of sensitive bacteria. One important group, the aminoglycosides, are bactericidal for reasons that are not understood, but the actions of these antibiotics may be due to their multiple effects on membranes, RNA metabolism and misreading of codons (triplets in mRNA (mRNA) composed of three nucleotides that represent a particular amino acid). The effects of this group depend on their concentrations and differ from drug to drug.

5.2 An outline of protein synthesis in bacteria

The linking of amino acids in specific sequences by peptide bonds takes place on the ribosomes. For this purpose, each amino acid is combined with its own specific transfer RNA (tRNA). Mupirocin (Section 5.9) prevents bacterial protein synthesis at this early stage rather than at the ribosome. It binds to the synthetase enzyme that forms isoleucyl-tRNA from isoleucine and its tRNA molecule and renders isoleucine unavailable for incorporation into proteins.

The instructions that determine the sequences of proteins are transcribed from the genes into a set of molecules of single-stranded mRNA. Ribosomes that are in the process of synthesizing protein are held

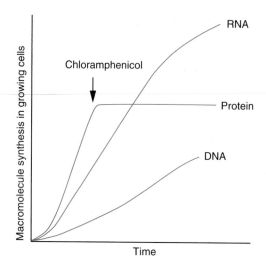

Figure 5.1: The effect of chloramphenicol on the synthesis of macromolecules in growing bacteria.

together in chains (polysomes) by the mRNA strands that specify the sequence of the protein. Ribosomes that are not engaged in protein synthesis are dissociated into their constituents (which in bacteria are the 30S and 50S components, equivalent to the 40S and 60S subunits of eukaryotic cells).

The usual model for protein synthesis involves two sites on the ribosome, one of which binds aminoacyl-tRNA (A site), while the other binds peptidyl-tRNA (P site). An alternative model envisages three binding sites [1] including an E or exit site, to which the deacylated tRNA binds before it detaches from the ribosome. This mechanistic difference has little bearing on the action of the many antibiotics that are effective at this site [2].

Protein synthesis begins with the assembly of the ribosome together with an mRNA molecule and the initiator species of tRNA that starts protein synthesis in bacteria, N-formylmethionyl-tRNA$_{metF}$. This species differs from methionyl-tRNA$_{metM}$ which is used to insert methionine in a position other than that of the initiation residue.

5.2.1 Initiation and the assembly of the ribosome

During the assembly of the initiation complex (*Figure 5.2*), the initiator tRNA is charged with methionine that is N-formylated in bacteria. This fmet-tRNA$_{metF}$ binds at the ribosome binding site (rbs) in the untranslated portion of the mRNA (a purine-rich sequence to the 5' end of the mRNA, not shown in *Figure 5.2*) and the smaller, 30S ribosomal

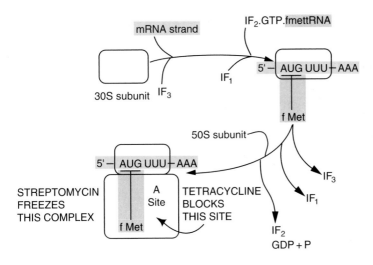

Figure 5.2: The assembly of the initiation complex for protein biosynthesis.

subunit. Three protein initiation factors are needed, of which IF_2 forms an aggregate with GTP and the fmet-tRNA$_{metF}$ molecule. The initiation complex is completed by the addition of the 50S ribosomal subunit and the release of three protein factors, IF_2 being released in association with GDP. At this stage, the peptidyl binding site (P site) of the 50S component is occupied by fmet-tRNA$_{metF}$ and the A site is vacant. The codon in mRNA opposite to the A site (in this example UUU, codes for phenylalanine) selects the aminoacyl-tRNA (phe-tRNA$_{phe}$ in this case) that can enter the A site, and therefore determines which amino acid will form the peptide bond. The tetracyclines and aminoglycosides are two families of antibiotics that are effective at this stage of the ribosomal cycle.

5.2.2 The ribosome cycle: entry of aminoacyl-tRNA

After the initiation complex has been assembled, the codon at the A site selects the aminoacyl-tRNA that can enter the A site next, and therefore determines which amino acid will be linked to formylmethionine by means of a peptide bond. The entry of aminoacyl-tRNA (*Figure 5.3*) to the A site resembles that of fmet-tRNA$_{metF}$ in that a complex is formed between the aminoacyl-tRNA, a protein factor (EF-Tu) and the nucleotide GTP. Elongation factor Tu is released with GDP produced by hydrolysis of GTP. The elongation factor Ts (EF-Ts), which is involved in reconverting EF-Tu:GDP back to EF-Tu:GTP by replacing the nucleoside triphosphate, is not shown in *Figure 5.3*.

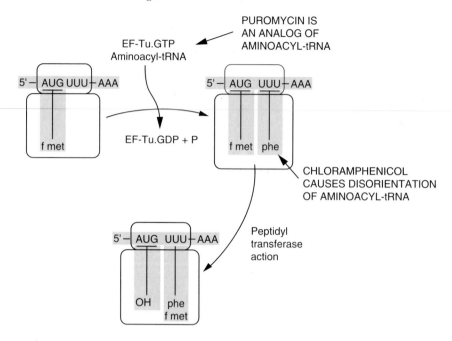

Figure 5.3: The entry of aminoacyl-tRNA and the synthesis of a peptide bond.

5.2.3 *The ribosome cycle: peptidyl transferase activity*

The entry of aminoacyl-tRNA to the A site completes the requirements for the formation of a peptide bond. The peptidyl transferase enzyme is a constituent of the 50S subunit of the ribosome, and it transfers formylmethionine (or the growing peptide in later cycles of synthesis of the protein) from the tRNA molecule in the P site. This residue is linked to the amino group of the amino acid (as shown in *Figure 5.4*) in the A site by a new peptide bond. Two antibiotics that affect this stage of synthesis are puromycin and chloramphenicol.

5.2.4 *The ribosome cycle: translocation of the ribosome*

The last stage of each cycle of synthesis is the translocation of the ribosome relative to the mRNA molecule. The translocation process requires the movement of the ribosome relative to the mRNA molecule, and during this process the tRNA molecule that has just released N-formylmethionine (or a peptide in later cycles) leaves the P site of the ribosome. The tRNA bearing the newly extended peptide enters the

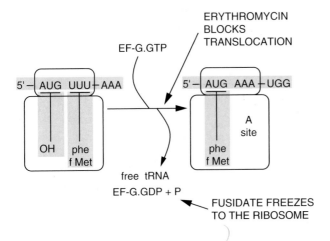

Figure 5.4: The translocation of the ribosome relative to mRNA.

P site leaving the A site vacant. A new codon (*Figure 5.4*) opposite the A site selects the next aminoacyl-tRNA that can enter this site. The process of translocation requires energy, which is supplied as GTP complexed with EF-G, another GTPase. The antibiotic erythromycin and other macrolides interfere with ribosome translocation. The steroid-like antibiotic fusidate does not prevent translocation, but does prevent the dissociation of EF-G from the ribosome after translocation has occurred.

The vacant ribosomal A site may now accept a new aminoacyl-tRNA, and the cycle of events described above is repeated for each new amino acid that is incorporated into the polypeptide chain. When the polypeptide has been completed, a 'stop' codon enters the ribosome opposite the A site. A termination factor probably binds to the stop codon and catalyzes the hydrolytic removal of the polypeptide from the tRNA molecule in the P site. The nascent polypeptide is thus released and the ribosome dissociates into its components, which are then available for reassembly to form a new initiation complex.

5.2.5 *Roles of ribosomal proteins and ribosomal RNA in the ribosome*

The ribosome is a complex in which both proteins and specific RNA molecules cooperate to bring together the required components for protein synthesis (*Figure 5.5*). The ribosomal RNA (rRNA) is of primary importance, and the proteins have the accessory role of modulating and stabilizing the functional sites on the ribosome [3]. Structural studies have indicated some of the interactions between these biomolecules.

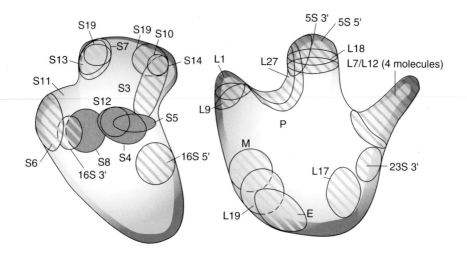

Figure 5.5: The location of proteins on the two subunits of the ribosome. Reproduced, with permission from the Annual Review of Biochemistry, Vol. 54, © 1985, by Annual Reviews Inc.

A loop of 23S RNA interacts with the peptidyl transferase enzyme of the large ribosomal subunit, and regions adjacent to this loop are concerned with tRNA binding at sites A, P and E [4]. Protein L23 is also involved with the A-site binding of aminoacyl-tRNA, and residue 2661 of 23S rRNA is particularly involved in binding the EF-Tu complex with GTP and aminoacyl-tRNA [5]. In 16S RNA, a set of nucleotides that are dispersed along the strand forms the 'mRNA-binding track' and the first 156 residues of 16S RNA form a domain that binds with near unit stoichiometry to four 30S subunit proteins. The ribosomal proteins and the rRNA are both involved in binding drugs that cause inhibition at the ribosomes [6].

During protein synthesis, the conformation of the ribosome may be altered by the loss of some rRNA helices and the formation of others. Knowledge of the roles of the ribosome components has developed alongside knowledge of the binding specificities of antibiotics for the ribosome constituents. Conserved sets of nucleotides in rRNA are protected from chemical attack in ribosomes by bound antibiotics [3], as well as by oligonucleotides that are complementary to the rRNAs. The antibiotic binding sites can be identified more precisely by such means.

5.3 Puromycin, a nonselective inhibitor

Although protein synthesis is similar in bacteria and eukaryotes, differences in detail permit some antibiotics to act selectively against bacteria. Puromycin (*Figure 5.6*) is an analog of aminoacyl-tRNA that is inhibitory

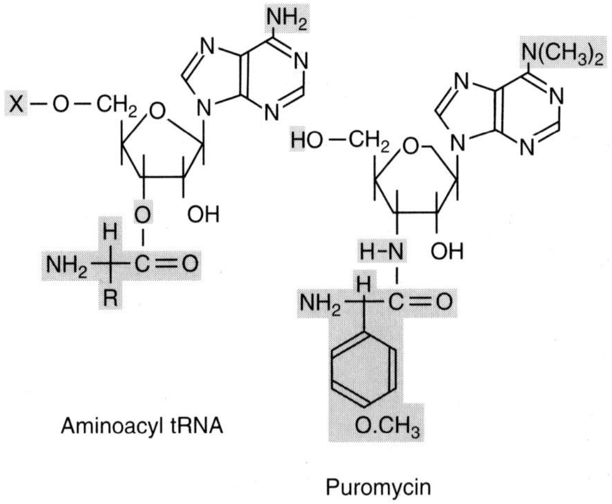

Figure 5.6: Puromycin, an analog of aminoacyl-tRNA.

in all protein-synthesizing systems. The nucleoside of puromycin is an analog of the terminal adenosine of tRNA to which amino acids are joined. Puromycin in the A site acts as an aminoacyl-tRNA analog in the peptidyl transferase reaction, and accepts the polypeptide chain, which is joined to the free amino group of the antibiotic. Further chain extension is prevented, and puromycin covalently linked to prematurely terminated polypeptide detaches from the ribosome. This reaction requires no energy input, indicating that the analogous formation of the peptide bond also requires no energy. As it is nonselective, puromycin has no clinical uses.

5.4 Streptomycin and other aminoglycosides

Streptomycin is effective against some Gram-positive bacteria, but was discovered in a survey conducted in order to find an antibiotic effective against the Gram-negative bacteria to complement the penicillins. Aminoglycosides consists of three (rarely two) carbohydrate rings linked by glycosidic bonds, and are strongly basic due to the possession of amino and guanido groups. The members of the streptomycin group have a guanido-containing streptidine ring, and other aminoglycosides having streptamine with amino groups instead (*Figure 5.7*) include the gentamicin group and the neomycin group. Amikacin, a semi-synthetic derivative of kanamycin, was developed as a poor substrate for the aminoglycoside-modifying enzymes for use against infections resistant to other aminoglycosides. Spectinomycin is not an aminoglycoside but contains an inositol residue with two methylamine substituents, but is often included in this group.

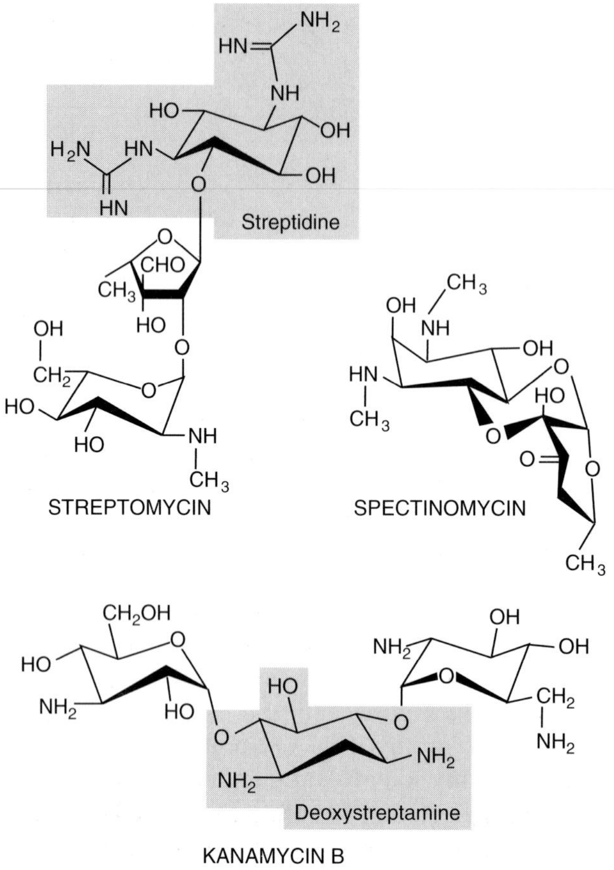

Figure 5.7: Aminoglycoside structures.

Streptomycin, the first member of the family, has been investigated in most detail, although it is no longer used in man. Its effects on sensitive micro-organisms include some membrane damage with loss of ions and metabolic intermediates and inhibition of respiration in cells with aerobic metabolism. It also causes increased RNA synthesis and mistranslation of the genetic message into proteins. Streptomycin-resistant mutants of *E. coli* differ from sensitive strains in their ribosomal proteins [2], and the critical effects of streptomycin are concerned with ribosome function. The synthesis of protein *in vitro* with components derived from strep[R] and strep[S] bacteria shows that the drug binds to the 30S subunit. Sensitivity is conferred by a small subunit from strep[S] cells with other components from strep[R] cells. The converse is true when only the 30S subunit is from strep[R] bacteria. In 30S particles reassembled from their components, the streptomycin-binding capacity resides mainly in S_{12} protein, which is mutated in strep^R cells. Mutations in at least nine other ribosomal proteins also contribute to streptomycin resistance. The S_{12} protein makes several intimate contacts with 16S rRNA, the proximal helix of which participates in streptomycin binding [7]. Protein synthesis *in vitro* with different concentrations of streptomycin indicates two alternative effects of the drug binding to the 30S subunit. A ribosome that is traversing mRNA and synthesizing a polypeptide will continue to do so (at a slower rate, and with incorporation errors) if streptomycin binds. Whole cells of *E. coli* exposed to low concentrations (2 μg ml^{-1}) of the drug synthesize protein molecules that contain incorrect amino acids due to inappropriate aminoacyl-tRNA insertions that do not match the codons at the A site in the distorted ribosome. This miscoding has a definite pattern in which U is misread as A or C and C is misread as A or U. Aminoglycosides vary in the extent of miscoding that they cause.

Cells treated with 20 μg ml^{-1} streptomycin cease to make protein, which, unlike a degree of miscoding, cannot be tolerated. The aminoglycosides bind to ribosomes so strongly that the ribosomes remain inhibited if the cells are extensively washed. Streptomycin incubated *in vitro* with polysomes (70S ribosomes linked by the mRNA that they are translating) causes the number of ribosomes attached to each mRNA to decrease to one, 'streptomycin monosomes', because the initiation complex ribosome cannot traverse the mRNA. Ribosomes, with or without bound streptomycin, that are already synthesizing polypeptides will continue to do so until completion and then separate from the mRNA. The blocked initiation complex ribosome is immobile, remains attached to the mRNA and denies other ribosomes access to the mRNA strand, which is degraded by ribonuclease.

The enzymes formed by resistant bacteria resemble the enzymes that protect the micro-organisms that produce these antibiotics (mostly *Streptomyces*), which are resistant to the antibiotics that they form, but not necessarily to different members of the same family.

5.4.1 Aminoglycoside administration

Aminoglycosides are not absorbed from the gut and are excreted in the feces. Deep intramuscular, subcutaneous or intrathecal injection may be used for systemic administration, but intravenous injection is the most rapid means of achieving effective concentrations. Some absorption may follow application to open wounds. Neomycin can be given by inhaled aerosol for the treatment of respiratory infections. The lipid solubility of aminoglycosides is low and they do not readily cross most membranes, but tend to remain in the extracellular fluid. Aminoglycosides may be administered as liposome preparations in order to overcome their lipid insolubility [8]. Streptomycin apparently crosses the placenta, but entry into the CSF is poor unless the meninges are inflamed. Low levels of entry into the cells limit aminoglycoside metabolism.

The plasma half-life of 2–4 h may increase in kidney failure as glomerular filtration is the major mechanism of excretion, and in such cases doses are reduced in order to avoid toxicity. The rate of excretion is reduced in infants because kidney function is not fully developed and toxic effects are more likely when aminoglycosides accumulate as a result of reduced clearance from the body.

5.4.2 Undesirable effects of the aminoglycosides

Hypersensitivity to aminoglycosides is independent of the dose, and occurs in about 5% of patients. It involves rashes and fever, and more rarely blood dyscrasias or anaphylactoid reactions. Dermatitis can occur after topical application. The pain of intramuscular injections may be reduced by including procaine in the injection as a local anaesthetic.

Damage to the 8th cranial nerve (ototoxicity) is proportional to the dose and duration of exposure. Labyrinthine damage causes headaches, dizziness, vertigo and loss of balance, some compensation occurs when the drug is withdrawn, but this can take several months. Auditory damage may result in total deafness, but early signs are loss of perception of high frequencies, and ringing in the ears. Neomycin, which is given orally to suppress the gut flora and to sterilize the bowel, has the most toxic effects on the auditory nerve, and treatment with this drug should not be prolonged. Inhalation of neomycin aerosol for treatment of lung infections has been known to cause deafness. Kanamycin is less ototoxic than neomycin, and under controlled conditions it can be given intramuscularly in order to treat septicemia.

Streptomycin administered to the peritoneum following surgery and neomycin applied near the diaphragm may cause respiratory paralysis as a result of neuromuscular blocking. Individuals with myasthenia gravis, or those treated with a muscle relaxant, are most likely to show such a response. Gentamicin is the most potent aminoglycoside, and shares many properties with streptomycin. It is given parenterally and excreted in the urine. The main undesirable effect of gentamicin is labyrinthine damage, the risk of which is greater than with kanamycin. The nephrotoxicity of gentamicin has not yet been firmly established.

5.4.3 Resistance to aminoglycosides

Resistance to aminoglycosides due to mutation of ribosomal proteins has been mentioned above because it cannot be separated from the

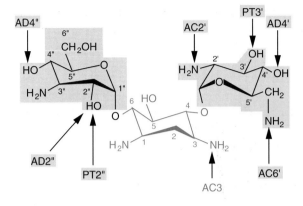

Figure 5.8: Inactivation of kanamycin by modifying enzymes.

description of the mode of action of these drugs. Significant resistance to aminoglycosides also involves the modification of these drugs (*Figure 5.8*) by diverse reactions that reduce their potency [9]. The enzymes responsible for such modification include acetyl transferases (AC), which use acetyl CoA to acetylate amino and hydroxyl groups, phosphotransferases (PT), which phosphorylate hydroxyl groups, and adenyl transferases (AD), which use ATP to derivatize hydroxyl groups with adenosine monophosphate residues. The enzymes that catalyze these reactions are abbreviated using the above letters together with numbers to show the position and a number of primes (') to indicate which ring is attacked (e.g. 3' O-phosphotransferase = PT3'). Because of resistance to aminoglycosides, the semi-synthetic compound amikacin was developed. Resistance to amikacin has been slow to develop in some environments, but a range of mechanisms now exists in diverse clinical isolates [10]. These include impermeability to the antibiotics, PT3', AC6' and AD4' in *P. aeruginosa* and AC6', PT2', AD4' and 4"O-AD in *S. aureus*.

5.5 The tetracyclines

These antibiotics are a similar group of compounds produced by *Streptomyces* (*Figure 5.9*) which vary somewhat in their pharmacology but resemble one another in antimicrobial activity. When resistance arises it is usually to all tetracyclines. Chlortetracycline has chlorine substituted at position 7 and oxytetracycline has a hydroxyl at position 5. At the plasma concentrations that can be achieved during chemotherapy, these drugs are bacteriostatic, although they

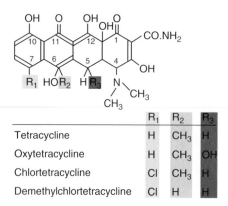

	R$_1$	R$_2$	R$_3$
Tetracycline	H	CH$_3$	H
Oxytetracycline	H	CH$_3$	OH
Chlortetracycline	Cl	CH$_3$	H
Demethylchlortetracycline	Cl	H	H

Figure 5.9: The structures of some tetracyclines.

are bactericidal at higher concentrations *in vitro*. The effects of tetracycline include the inhibition of transport across membranes, and the inhibition of the metabolism of glucose and of oxidative phosphorylation, but the common tetracyclines inhibit protein synthesis. Tetracycline, chlortetracycline, 6-demethyl-6-deoxytetracycline, minocycline and doxycycline inhibit cell-free translation of mRNA. They bind to the 30S subunit at a site consisting of proteins S7, S14, S19 and the 3' domain of 16S RNA [11]. One molecule of tetracycline per ribosome causes complete inhibition. Tetracyclines complex with divalent metal ions, and ribosome binding may involve magnesium ions, forming a rigid complex.

Polyphenylalanine synthesis directed by polyuridylate (U) in a cell-free system is less sensitive to tetracyclines on ribosomes derived from a tet^R strain. The 50S subunit plays no part in the inhibition. When ribosomes are mixed with synthetic mRNA, tetracyclines inhibit the binding of amino-acyl-tRNA at the A site, but not of peptidyl-tRNA that would be expected to bind at the P site.

Tetracycline selectivity against bacteria does not occur at the ribosome level because *in vitro* bacterial (70S) or eukaryotic (80S) ribosomes are both inhibited. Selectivity takes place at the membrane, because tetracyclines are actively accumulated by bacteria using an energy-dependent process, and one method whereby resistance develops is the reversal of tetracycline-concentrating activity. Eukaryotic cells do not concentrate tetracyclines and are little affected by the usual therapeutic doses, but treatment with drugs that alter membrane permeability may facilitate the entry of tetracyclines into eukaryotic cells, and so cause an anti-anabolic effect.

Tetracyclines inhibit the matrix metalloproteinases such as collagenase [12] that are implicated in connective tissue degradation, and in arthritis these enzymes are suppressed by oral tetracycline. Metalloproteinases of the periodontium are inhibited by tetracyclines with clinical benefit to the progress of periodontal disease. Tetracyclines may therefore be useful in the prevention of tissue damage in chronic inflammatory conditions [13].

A second class of tetracyclines [14, 15] discovered more recently and including anhydrotetracycline and anhydrochlortetracycline, do not inhibit protein synthesis. In contrast to the classical tetracyclines, their primary action is the lysis of the cell membrane in *E. coli*. Electron microscopy shows cells with morphological alterations, cell ghosts and cell debris. This class of tetracycline may interfere with the membrane's electrochemical gradient, which in turn leads to stimulation of autolytic enzyme activity.

5.5.1 Tetracycline administration

Tetracyclines are basic and rather insoluble, but form soluble sodium salts, and may be administered orally, parenterally or topically. Absorption from the gut is variable in different individuals, but often produces effective plasma concentrations. Chlortetracycline, tetracycline, oxytetracycline and methacycline are not so well absorbed as doxycycline and minocycline, and food in the gut can further decrease absorption. Chelation of calcium ions causes precipitation, and iron salts and antacids such as aluminum hydroxide and sodium bicarbonate also decrease absorption. Phosphates decrease free calcium levels in the lumen of the gut and facilitate tetracycline absorption. In order to avoid the uncertainty of absorption, tetracyclines are often administered intravenously.

Tetracyclines bind to varying degrees to plasma proteins, but there is sufficient unbound drug to ensure entry to most tissues. The drugs are excreted in the bile, and some reabsorption may occur. In CSF the concentration may increase to about 25% of the plasma value. Because of irritation, tetracyclines must not be given intrathecally.

Tetracyclines are used in the treatment of acne, soft tissue bacterial infections, Lyme disease (borreliosis), chlamydia and respiratory tract infections. The second-generation drugs minocycline and doxycycline are widely prescribed, and their pharmacological and microbiological characteristics offer advantages for application in dermatology [16].

5.5.2 Undesirable effects of tetracyclines

Tetracyclines are deposited in developing teeth and bones, presumably as calcium complexes, and the discoloration of teeth persists, so children below 8 years of age are not treated with these antibiotics. Tetracyclines also cross the placenta and should not be given to pregnant women. Skin rashes and urticaria may occur in sensitive individuals after treatment with any tetracycline, and cross-sensitivity is general. Allergic reactions of the anaphylactoid type can occur, but gastrointestinal disturbances such as nausea, vomiting and pain are usually dose-dependent. Diarrhea free of blood and leukocytes may be caused by irritation of the mucosa. Super-infection of the gut following the suppression of the normal bacterial population may cause diarrhea with blood and leukocytes in watery stools. Under such circumstances it is necessary to withdraw the tetracycline, identify the micro-organism responsible for the super-infection, suppress it with an alternative antibiotic, and restore fluid and ionic balance by intravenous fluid infusion. Tetracyclines may be anti-anabolic causing weight loss and a negative nitrogen balance by reducing protein synthesis, and they should not be given to malnourished patients. There may be deterioration of liver and kidney function following tetracycline administration in excess of 2 g daily. Liver damage involves the development of fatty deposits, jaundice and a high rate of nitrogen excretion. The livers of pregnant women appear to be especially sensitive, but there is no evidence of teratogenic effects.

5.5.3 Resistance to tetracyclines

Resistance to tetracyclines has received much attention [17–20] and may reside in the cell membrane, in mutational changes at the the ribosome, or in the destruction of the drug (*Figure 5.10*). The genes for resistance may be located on plasmids, the chromosome or on the mobile chromosomal elements known as transposons.

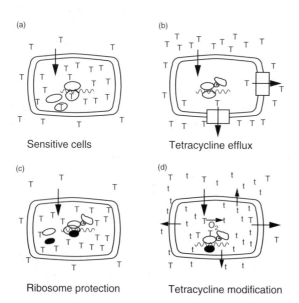

(a)

Sensitive cells

(b)

Tetracycline efflux

(c)

Ribosome protection

(d)

Tetracycline modification

Figure 5.10: Mechanisms of resistance to tetracyclines. Reproduced from Salyes *et al.* (1990) *Molecular Biology* 4(1) 151–156. With permission from Plenum Publishing Corporation.

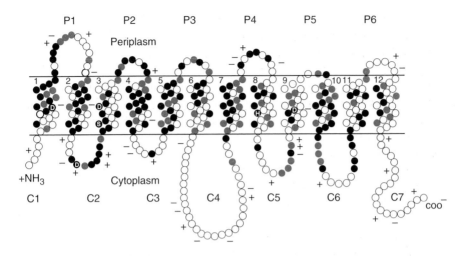

Figure 5.11: Tetracycline transport protein TetA. Reproduced with permission from the American Society of Microbiology from ref. 21, Allard, J.D. and Bertrand K.P., *Journal of Bacteriology,* Vol. 175, pp. 4554–4560. © 1993.

The resistance gene *tetA* is used as a genetic marker in the vector pBR322. It encodes a 41-kDa inner membrane protein that is a tetra-cycline/H$^+$ antiporter (i.e. it transports a proton in the opposite direction to tetracycline). The set of proteins that cause tetracycline export (TetA to TetE, TetG, TetK, TetL) are homologous [21], and the topology of these enzymes consists of 12 transmembrane segments (TM1, etc.) with up to five periplasmic loop segments and both the termini (C1, C7), together with five further loops (C2, etc.) in the cytoplasm (*Figure 5.11*). Highly conserved residues known to be related to function in TetE efflux protein [21] are all hydrophilic and predominantly ionized residues. They occur in the transmembrane segments together with the C2 loop, and consist of four aspartic acid (D) residues, one arginine (R) residue, one histidine (H) residue and one serine (S) residue.

In the TetB membrane transport protein [22], Asp15 within the first predicted transmembrane helix is conserved in all Gram-negative bacteria. Its replacement lowered the V_{max} of the tetracycline efflux system by 90%. In tetracycline/H$^+$ antiporters the region including Ser65–Asp66 acts as a gate. The C2 interhelix also contains a sequence motif of 11 residues found in all Tet resistance proteins. A significant decrease in transport is associated with replacement of either the glycine residue of the loop or the one that immediately precedes it, perhaps due to steric hindrance by other larger side-chains on the β-turn struc-ture. The polycationic nature of the loop is not critical for activity, but the replacement of Arg70 (R in the C2 loop) (*Figure 5.11*) causes complete loss of the activity except when it is replaced by lysine, so a positive charge at this position is required. It is essential to have a negative charge in loop C2 at Asp66 (D in loop C2) (*Figure 5.11*).

A means of resistance to tetracyclines sited at the ribosome has been found in various species. The ribosomal genes *tetM* (which is the commonest determinant of resistance [23]) and *tetO* share more than 75% amino-acid-sequence identity. Resistance to tetracycline and minocycline was due to the *tetM* gene in *Listeria* associated with the movements of Tn1545-like conjugative transposons. Cotransfer of TetM occurred at frequencies similar to those observed with Tn1545. The *Bacteroides tetQ* gene has a deduced amino acid sequence that is only 40% identical to *tetM* or *tetO*. This resistance gene has emerged recently and has spread only within the *Bacteroides* group [24].

The *tetL* resistance gene, the second most prevalent tet^R gene in enterococci and streptococci is borne by a 5-kb plasmid. The gene from *Staphylococcus hyicus* is induced by tetracycline and encodes a hydrophobic protein composed of 458 amino acids [25]. Thirty amino acids are conserved among variants of this protein, which is closely related to the TetL proteins of small plasmids in *Bacillus*. There is a large evolutionary distance between these and TetM from *Streptococcus faecalis* and *Staphylococcus aureus*, and TetO from *Streptococcus mutans*. The movable genetic elements, namely transposons and plasmids, are responsible for the transfer of tetracycline resistance from enterococci and streptococci to *Listeria monocytogenes* [26]. Some bacteria detoxify tetracyclines. The TetF determinant of *Bacteroides fragilis* is of this type, and neither tetracycline nor its degradation products are transported out of the cell [27].

5.6 Chloramphenicol

Originally isolated from a streptomycete, chloramphenicol (chloromycetin) is a simple drug (*Figure 5.12*) that can be synthesized cheaply and has a broad range of effectiveness. One analog of chloramphenicol, thiamphenicol, has been used. Of the four isomeric forms (D and L, threo and erythro), only D-threo-chloramphenicol is effective. It stops protein synthesis rapidly without affecting DNA or RNA synthesis, and it does not bind to eukaryotic 80S ribosomes, but its effectiveness against protein synthesis on 70S ribosomes causes problems in rapidly dividing tissues. Mitochondria are probably derived from symbiotic bacteria and mitochondrial ribosomes are 70S, and sensitive to chloramphenicol. The reproduction of mitochondria requires proteins coded by mitochondrial DNA to be synthesized on mitochondrial ribosomes.

The drug binds relatively weakly to the 50S particle without affecting the 30S subunit. It inhibits the formation of peptide bonds without blocking translocation of mRNA, and it inhibits the puromycin reaction in intact bacteria. The normal binding of aminoacyl-tRNA in the A site

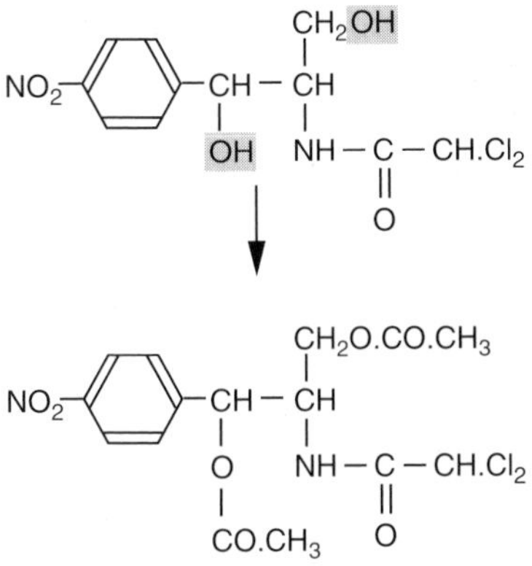

Figure 5.12: D-threo-chloramphenicol and its modification by acetylation.

is disrupted by chloramphenicol, so that peptidyl transferase cannot operate. This was demonstrated using aminoacyl-tRNA fragments, from which much of the tRNA was removed by digestion with ribonuclease, leaving only the fragment close to the amino acid intact. These aminoacyl-tRNA fragments bind only weakly to the ribosomes and this binding is specifically prevented by the microbiologically active isomer D-threo-chloramphenicol, but not by the other three isomers. Chloramphenicol therefore affects the orientation of the part of aminoacyl-tRNA that is close to the amino acid residue in a way that prevents peptide bond formation.

Chloramphenicol is a broad spectrum antibiotic that is effective against Gram-positive and Gram-negative bacteria as well as mycoplasmas, rickettsia and chlamydia, but not against pseudomonas. However, because of its toxicity its use is limited to severe infections that do not respond to other antibiotics. It is still used against typhoid fever (*Salmonella typhi*) if it is chloramphenicol sensitive. Amongst cases reported over a period of 5 years, *S. typhi* was sensitive to chloramphenicol in all the cases where it was cultured. In developing countries bacterial meningitis is associated with a high fatality rate and injection of chloramphenicol is valuable as a first-line presumptive treatment for meningitis.

5.6.1 Chloramphenicol administration

Chloramphenicol 'micro-fined' powder is well absorbed from the gut. The palmitate ester is given orally and is hydrolyzed before absorption takes place, and chloramphenicol sodium succinate is used for intravenous injection. Chloramphenicol penetrates all tissue compartments, and is effective in the CSF and against intracellular bacteria.

Conjugation in the liver forms glucuronides that are not bactericidal. Metabolism is faster if liver microsomal enzymes have been induced by drugs such as phenobarbitone. Chloramphenicol is filtered at the glomerulus, and the glucuronide is secreted by the tubules so accumulation may occur if kidney function is impaired.

5.6.2 Undesirable effects of chloramphenicol

Hypersensitivity reactions include skin rashes, fever, gastrointestinal discomfort and soreness of the mouth. In newly born babies, chloramphenicol induces the 'gray syndrome' of symptoms which include failure to feed, hypothermia, muscular flaccidity and an ashen gray pallor. Gray syndrome, which is fatal in about half of all cases, is associated with the inability to conjugate and excrete chloramphenicol.

Chloramphenicol may also cause suppression of development of blood cells made in the bone marrow. This hypersensitivity response is often fatal, and may occur months after treatment in about 1 in 40 000 patients. Because this effect is not dose-related, there is now some concern even about the topical use of chloramphenicol to treat conjunctivitis. Reversible dose-dependent depression of the bone marrow, leading to reticulocytopenia and anemia, is not associated with total suppression of bone-marrow cells, and is much more common than the disastrous aplasia. Optic neuritis, with loss of visual acuity, is generally reversible. Administration of vitamins of the B complex is reported to speed the recovery of visual acuity.

5.6.3 Resistance to chloramphenicol

Resistance to chloramphenicol occurs in both Gram-negative and Gram-positive bacteria, and is transmissible in the Gram-negative enteric genera. Plasmid-specified chloramphenicol acetyltransferase (CAT) catalyzes the transfer of an acetyl group from acetyl-CoA to both hydroxyl groups on the drug, thus inactivating it (*Figure 5.12*). The kinetics show that the K_m increases (i.e. the affinity decreases) for the acetyl acceptors in the drug and also for the donor, acetyl-CoA in several resistant mutants. The binding of CAT protein to chloramphenicol and its acetyl derivatives has been investigated in detail by X-ray diffraction of crystals [28].

5.7 The macrolides

5.7.1 Erythromycin

The macrolide antibiotics have structures that contain large (12–16 atoms) lactone ring systems, and carbohydrate side-chains that include

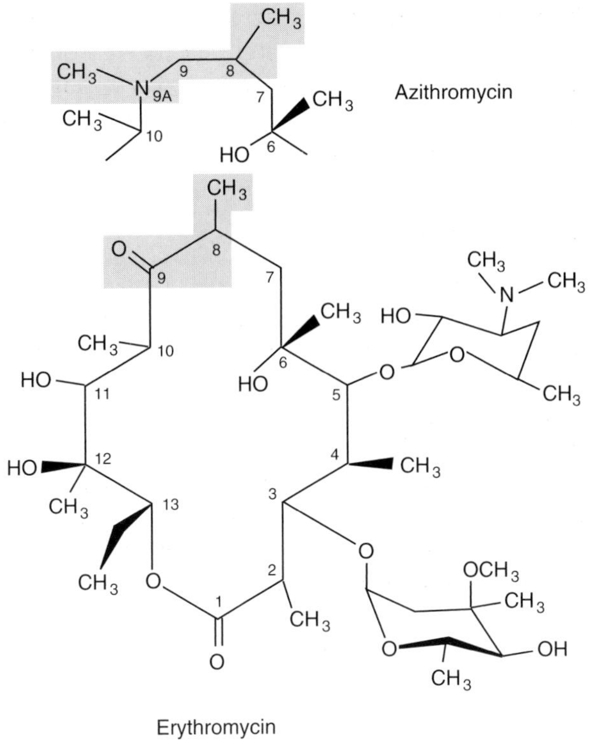

Figure 5.13: Structure of the macrolides: erythromycin and azithromycin.

amino sugars. The macrolides include spiramycin, oleandomycin, lankamycin and carbomycin, but the most important macrolide clinically is erythromycin (*Figure 5.13*). Azithromycin, clarithromycin and roxithromycin are recent members of the group [29, 30]. Macrolides inhibit protein synthesis in bacteria and in cell-free systems. Erythromycin binds to the 50S ribosomal subunit of sensitive bacteria but not to that of resistant mutants. The effect of erythromycin may be exerted on the translocation of ribosomes, and it causes the release of incomplete polypeptides from the ribosomes.

The antibacterial spectrum of this group of antibiotics is similar to that of the penicillins, namely most effective against the Gram-positive cocci and many Gram-positive rods. Gram-negative bacteria only achieve about 1% of the intracellular concentration of the drug that occurs in Gram-positive strains. Its main use is against staphylococcal, streptococcal and pneumococcal infections, but it is the drug of first choice for treatment of meningitis caused by *Flavobacterium meningosepticum*, a Gram-negative rod. It is a drug of second choice for most Gram-positive and some Gram-negative bacteria, as well as for some mycoplasmas and chlamydia. Azithromycin, a macrolide of a new class called azalides, is highly effective at a concentration of 2 μg ml⁻¹ in the treatment of *A. actinomycetemcomitans*, against which erythromycin shows poor activity. Since azithromycin has favorable pharmacokinetics, including excellent distribution into tissues, it could be expected to pass into the gingival crevicular fluid to treat *A. actinomycetemcomitans*-associated periodontitis [31].

5.7.2 Macrolide administration

Erythromycin, the most important and widely used member of this group, is destroyed by stomach acid. Coated pills and the estolate ester are administered orally, but the latter is not active until the estolate has been hydrolyzed (mainly in the liver) to produce free erythromycin. Erythromycin glucoheptonate can be given intravenously in order to establish an effective plasma concentration rapidly. Erythromycin is distributed in most of the body water compartments but penetration into the CSF is poor unless the meninges are inflamed.

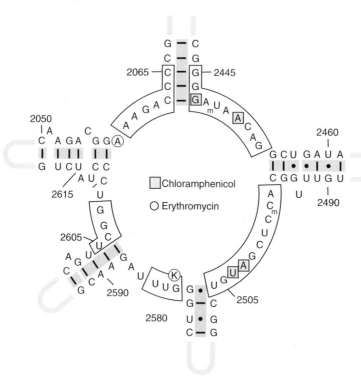

Figure 5.14: The region of 23S rRNA concerned with chloramphenicol and erythromycin resistance.

Presumably some erythromycin is metabolized to form compounds which have not been identified. Biliary secretion is the major route of elimination in man, and there may be some reabsoption via the entero-hepatic circulation.

5.7.3 Macrolide toxicity

Erythromycin base is not toxic, but side-effects associated with the use of erythromycin estolate include abdominal pain, fever, enlargement of the liver and decreased biliary secretion (cholestatic hepatitis), and these most frequently occur after 2 weeks of treatment. Such side-effects are attributed to hypersensitivity to estolate, and although recovery is satis-factory, recurrence is likely. Nausea, vomiting, diarrhea and allergic skin reactions also occur in a small percentage of patients.

5.7.4 Resistance to macrolides

Resistance is usually caused by a methylase of 23S rRNA that may be specified by plasmids or transposons [30]. An adenosine residue at posi-tion 2058 of the *E. coli* sequence in the peptidyl transferase domain is methylated once or twice (*Figure 5.14*) by this enzyme.

5.8 Fusidic acid

This antibiotic is unusual in that it has a structure which resembles the steroids (*Figure 5.15*). It is usually employed as the more soluble sodium salt. Fusidate is effective against Gram-positive but not against Gram-negative bacteria, although it will inhibit protein synthesis in cell-free preparations of Gram-negative bacteria. Translocation of the mRNA is not inhibited, but the direct effect of fusidate appears to involve binding of the complex of elongation factor EF-G and GDP to the ribo-some after the translocation event [32]. Hydrolysis of excess GTP by EF-G in the presence of functional ribosomes is halted by fusidate. EF-G and EF-Tu have overlapping sites on the ribosome (*Figure 5.16*).

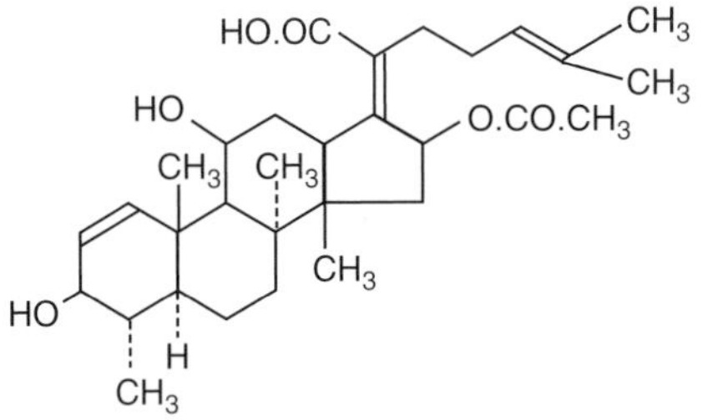

Figure 5.15: Fusidic acid structure.

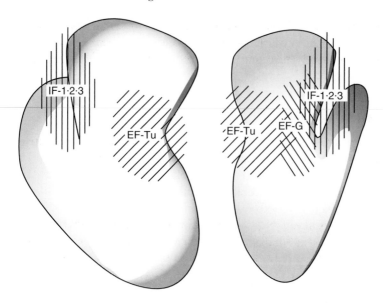

Figure 5.16: Binding sites for initiation and elongation factors on the 30S subunit of the ribosome.

The retained EF-G renders the EF-Tu site unavailable for binding the aminoacyl-tRNA/EF-Tu/GTP. EF-G from fus[R] bacteria is not bound to the ribosome by fusidate in this way, and if EF-G from such a strain is used in an *in vitro* system, protein synthesis is not inhibited by fusidate. Fusidic acid, which is usually given concurrently with another antibiotic, is effective against a wide variety of staphylococcal infections, including acute osteomyelitis and septic arthritis [33].

5.8.1 Fusidate administration

This compound is well absorbed when taken orally, and readily penetrates most tissues apart from the CSF. It can also be given intravenously as the diethanolamine salt. Excretion is mainly by biliary secretion of the glucuronide conjugate, although some metabolites are excreted in the urine. Ophthalmic complaints needing topical antibiotic treatment may be treated with fucithalmic as an alternative to chloramphenicol. Fusidic acid is notable for its activity against staphylococci, whether or not they are resistant to methicillin (MRSA). It is also active against many genera of strict anaerobes and microaerophiles. A combination of fusidate with other antibiotics generally gives enhanced action or no change in effect, but it also delays the emergence of resistant mutants. Some antagonism of action has been reported when fusidate is tested with penicillins *in vitro*, presumably because fusidate prevents growth, whereas penicillins are only effective against growing cells. The clinical significance of this *in vitro* antagonism is difficult to evaluate, but complementation of antibacterial action has been reported when fusidate and penicillins are used *in vivo*.

5.8.2 Undesirable effects of fusidate

Toxic effects are mild gastrointestinal upset and mild skin rashes occurring in some patients, but oral therapy is generally well tolerated.

5.8.3 Resistance to fusidate

Staphylococcus aureus resistance due to a one-step chromosomal mutation is readily demonstrated *in vitro*. Such mutants appear to be defective, and are infrequently encountered clinically. Most isolates from patients have plasmid-mediated resistance. The risk of resistance emerging is low when fusidic acid is used to treat acute infections, but higher in chronic infections. Over a 20-year period of use of fusidate to manage staphylococcal infection, the frequency of resistance has remained below 2% [34]. Mutants showing resistance by more than one mechanism may readily be selected *in vitro*, and fus^R mutants of EF-G may show altered interaction with GTP.

5.9 Mupirocin or pseudomonic acid

Mupirocin (bactroban) (*Figure 5.17*) prevents bacterial protein synthesis, but is unusual in not being effective at the ribosome. It binds to the synthetase that combines isoleucine with its tRNA and so makes this amino acid unavailable for incorporation into proteins at the ribosome, which will stall at codons for isoleucine. Its activity is limited to some Gram-positive aerobes, and as it is rapidly hydrolyzed if given systemically. It is used in creams and ointments for topical treatment of staphylococci and pyogenic streptococci (group A) in impetigo [35], as an alternative to systemic antibiotics such as erythromycin, β-lactams and clindamycin [36]. It is effective against MRSA, and mupirocin ointment has been used to decrease the carriage of staphylococci in the anterior nares, but it seems preferable to save it for use in outbreaks of MRSA infection [37]. Resistance to mupirocin is often plasmid borne, but at least one case of *S. aureus* resistance involves a novel gene for the synthetase [38] that is distinct from the native *ileS* gene.

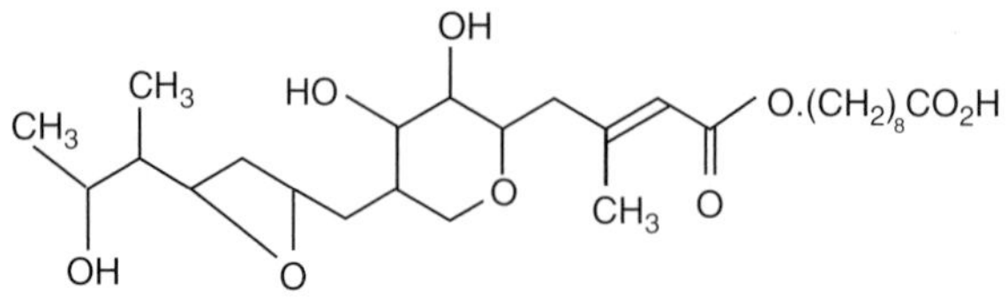

Mupirocin

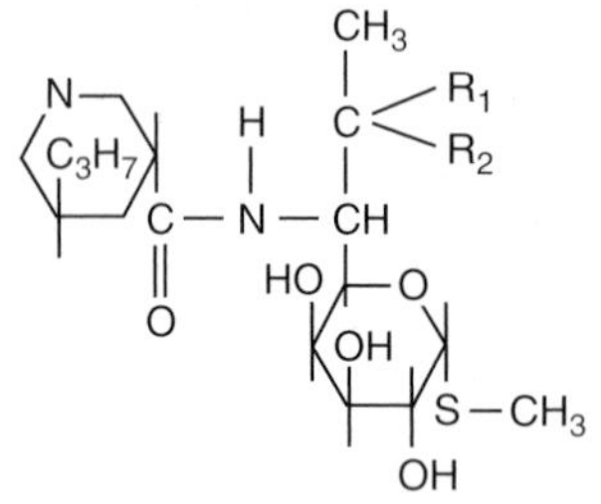

Lincomycin R$_1$ = -OH R$_2$ = -H
Clindamycin R$_1$ = -H   R$_2$ = -Cl

Figure 5.17: Structures of mupirocin, clindamycin and lincomycin.

5.10 Clindamycin and lincomycin

Clindamycin is a less toxic derivative of lincomycin (*Figure 5.17*), and it has a similar spectrum of activity to erythromycin. The site of action is at the 50S subunit of the ribosome, where the binding site overlaps with those of erythromycin and chloramphenicol. The binding of all three antibiotics depends on protein L16.

5.11 MRSA

Molecular and genetic analysis of multiresistant isolates of *S. aureus* from widely separated hospitals shows that resistance may be encoded in different ways (*Table 5.1*). Many strains contain small chloramphenicol resistance plasmids and also a family of much larger multiresistance plasmids that encode resistance to trimethoprim and penicillin as well as to the aminoglycosides. On these plasmids, transposons code for the 57-kDa bifunctional protein that has both aminoglycoside acetyltransferase [AT(6')] and phosphotransferase [PT(2")] activities. Transposition and genetic rearrangement have also contributed to the evolution of a multiresistant chromosome in this micro-organism, in the majority of which are encoded the determinants for resistance to a wide range of antibiotics including erythromycin, fusidic acid, methicillin, minocycline, rifampicin, spectinomycin, streptomycin, sulfonamides, tetracycline and a β-lactamase. Some of these chromosomal resistance determinants were plasmid encoded prior to 1970 [39].

Table 5.1: Resistance mechanisms in MRSA to antibiotics effective against protein synthesis

Antibiotic	Resistance gene[a]	Mechanism of resistance
Mupirocin	C	Unknown
Amikacin, kanamycin and gentamicin, tobramycin	P C T IS	6' N-acetylation 2" O-phosphorylation
Above except for gentamicin	P C IS	4'4" O-adenylation
Tetracyclines	P C IS	Efflux from cells (TetK, TetL)
	P	Ribosome mutation (TetM)
Chloramphenicol	P	Acetylation
Erythromycin	P C T	rRNA methylation
Clindamycin	P C T	rRNA methylation

[a]C, chromosomal; P, plasmid; T, transposon; IS, insertion sequence.

References

1. Hausner, T.-P., Geigenmuller, U. and Nierhaus, K.H. (1988) *J. Biol. Chem.*, **263**, 13103–13111.
2. Cannon, M. (1990) in *Comprehensive Medicinal Chemistry*, Vol. 2 (C. Hansch, P.G. Sammes and J.B. Taylor, eds). Pergamon, Oxford, pp. 813–838.
3. Dahlberg, A.E. (1989) *Cell*, **57**, 525–529.
4. Moazed, D. and Noller, H.F. (1989) *Cell*, **57**, 585–597.
5. Tapio, S. and Isaksson, L.A. (1991) *Eur. J. Biochem.*, **202**, 981–984.
6. Moazed, D. and Noller, H.F. (1987) *Nature*, **327**, 389–394.
7. Pinard R., Payant, C., Melancon, P. and Brakier-Gingras, L. (1993) *FASEB J.*, **7**, 173–176.
8. Karlowsky, J.A. and Zhanel, G.G. (1992) *Clin. Infect. Dis.*, **15**, 654–667.
9. Shaw, K.J., Rather, P.N., Hare, R.S. and Miller, G.H. (1993) *Microbiol. Rev.*, **57,** 138–163.
10. Jackoby, G.A. and Archer, G.L. (1991) *N. Engl. J. Med.*, **324**, 601–612.
11. Weiner, L. and Brimacombe, R. (1987) *Nucleic Acids Res.*, **15**, 3635–3670.
12. Greenwald, R.A., Moak, S.A., Ramamurthy, N.S. and Golub, L.M. (1992) *J. Rheumatol.*, **19**, 927–938.
13. Kinane, D.F. (1992) *Curr. Opin. Dent.*, **2**, 25–32.
14. Maibach, H. (1991) *Cutis*, **48**, 411–417.
15. Rasmussen, B., Noller, H.F., Daubresse, G. *et al.* (1991) *Antimicrob. Agents Chemother.*, **35**, 2306–2311.
16. Oliva, B., Gordon, G., McNicholas, P., Ellestad, G. and Chopra, I. (1992) *Antimicrob. Agents Chemother.*, **36**, 913–919.
17. Salyers, A.A., Speer, B.S. and Shoemaker, N.B. (1990) *Mol. Microbiol.*, **4**, 151–156.
18. Chopra, I., Hawkey, P.M. and Hinton, M. (1992) *J. Antimicrob. Chemother.*, **29**, 245–277.
19. Levy, S.B. (1992) *Antimicrob. Agents Chemother.*, **36**, 695–703.
20. Speer, B.S., Shoemaker, N.B. and Salyers, A.A. (1992) *Clin. Microbiol. Rev.*, **5**, 387–399.
21. Allard, J.D. and Bertrand, K.P. (1993) *J. Bacteriol.*, **175**, 4554–4560.

22. Burdett, V. (1991) *J. Biol. Chem.*, **266**, 2872–2877.
23. McMurry, L.M., Stephan, M. and Levy, S.B. (1992) *J. Bacteriol.*, **174**, 6294–6297.
24. Nikolich, M.P., Shoemaker, N.B. and Salyers, A.A. (1992) *Antimicrob. Agents Chemother.*, **36**, 1005–1012.
25. Schwarz, S., Cardoso, M. and Wegener, H.C. (1992) *Antimicrob. Agents Chemother.*, **36**, 580–588.
26. Poyart-Salmeron, C., Trieu-Cuot, P., Carlier, C., MacGowan, A., McLauchlin, J. and Courvalin, P. (1992) *Antimicrob. Agents Chemother.*, **36**, 463–466.
27. Park, B.H. and Levy, S.B. (1988) *Antimicrob. Agents Chemother.*, **32**, 1797–1800.
28. Murray, I.A., Lewendon, A., Williams, J.A., Cullis, P.M., Shaw, W.V. and Leslie, A.G. (1991) *Biochemistry*, **30**, 3763–3770.
29. Kirst, H.A. and Sides, G.D. (1989) *Antimicrob. Agents Chemother.*, **33**, 1413–1422.
30. Ballow, C.H. and Amsden, G.W. (1992) *Ann. Pharmacother.*, **26**, 1253–1261.
31. Pajukanta, R., Asikainen, S., Saarela, M., Alaluusua, S. and Jousimies-Somer, H. (1992) *Antimicrob. Agents Chemother.*, **36**, 1241–1243.
32. Verbist, L. (1990) *J. Antimicrob. Chemother.*, **25** (Suppl. B), 1–5.
33. Coombs, R.R. (1990) *J. Antimicrob. Chemother.*, **25** (Suppl. B), 53–60.
34. Shanson, D.C. (1990) *J. Antimicrob. Chemother.*, **25** (Suppl. B), 15–21.
35. Goldfarb, J., Crenshaw, D., O'Horo, J., Lemon, E. and Blumer, J.L. (1988) *Antimicrob. Agents Chemother.*, **32**, 1780.
36. Dagan, R. and Bar-David, Y. (1992) *Antimicrob. Agents Chemother.*, **36**, 287.
37. Kaufman, C.A., Terpenning, M.S., Xiaogong, H.E., Zarins, L.T., Ramsey, M.A., Jorgensen, R.N., Sottile, W.S. and Bradley, S.F. (1993) *Am. J. Med.*, **94**, 371.
38. Hodgson, J.E., Curnock, S.P., Dyke, K.G., Morris, R., Sylvester, D.R. and Gross, M.S. (1994) *Antimicrob. Agents Chemother.*, **38**, 1205–1208.
39. Skurray, R.A., Rouch, D.A., Lyon, B.R., Gillespie, M.T., Tennent, J.M., Byrne, M.E., Messerotti, L.J. and May, J.W. (1988) *J. Antimicrob. Chemother.*, **21** (Suppl. C), 19–39.

Chapter 6

Membrane-active drugs effective against bacteria and fungi

6.1 The cytoplasmic membrane as a target for selective antimicrobial action

The cytoplasmic membrane of microbial cells has the same basic structure as that of higher cells. It appears in thin sections under the electron microscope as a double-layered structure around 80 Å thick. Bacterial membranes contain 50–70% protein and 20–30% phospholipid, with other glycolipids making up the remainder. The phospholipids form a bilayer, with the hydrophilic head groups on the outer and inner faces and the fatty acids providing a hydrophobic core. The membrane proteins are distributed within the bilayer; some are exposed on either the inner or outer faces, while others span the whole bilayer structure. The proteins have important functions involving transport of nutrients, generation of energy, synthesis and assembly of cell-wall components, export of cellular products and monitoring of the external environment. The stability, integrity and functioning of the cytoplasmic membrane depend upon the maintenance of a range of noncovalent interactions between the protein and lipid components. These involve hydrophobic (van der Waals) forces between the alkyl chains of the lipids, ionic interactions between polar groups on lipids and associated divalent metal ions, and hydrogen bonding between proteins and lipids. A key feature that distinguishes bacteria from eukaryotic organisms (fungi and protozoa) and mammalian cells is the absence of sterols in bacterial membranes. In fungi, the membrane contains molecules of ergosterol which are located between adjacent phospholipids in the lipid bilayer and account for around 5% of the membrane. Mammalian cells also contain a sterol component, but it is cholesterol instead of ergosterol.

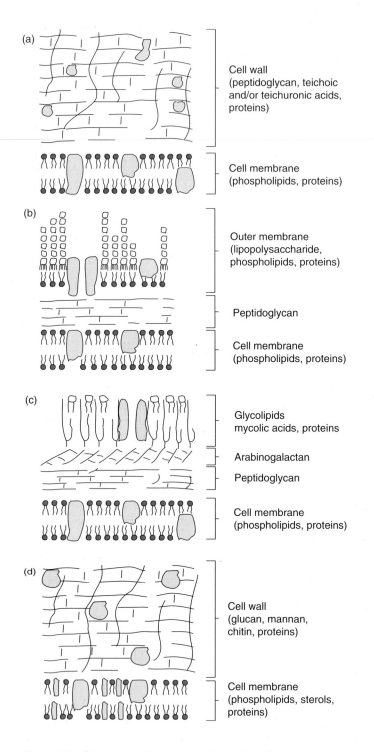

Figure 6.1: Schematic diagram of the wall and membrane structures surrounding: (a) Gram-positive bacteria; (b) Gram-negative bacteria; (c) mycobacteria and (d) fungi.

Most microbial cells have a thick wall structure surrounding the cytoplasmic membrane (*Figure 6.1*), providing it with mechanical support against the high internal osmotic pressure of the cells (see Section 3.2). In Gram-negative bacteria there is an additional membrane structure, the outer membrane, which covers the outer surface of the cell wall. This has a bilayer structure composed of proteins and lipids, but the lipid components are distributed asymmetrically. Phospholipids are located on the inner face and lipopolysaccharides (glycolipids linked to polysaccharide chains of varying lengths) are located on the outer face, with the polysaccharide chains protruding outward from the cells. Mycobacteria have an even more complex wall structure. The outer layers contain an arabinogalactan polysaccharide in addition to peptidoglycan, plus a variety of high-molecular-weight lipids, including the mycolic acids that contain long hydrocarbon chains. The outer membrane of Gram-negative bacteria and the lipid-rich wall of mycobacteria present a significant barrier to the penetration of antimicrobial agents. Small hydrophilic nutrients can cross the barrier via aqueous channels provided by the porin proteins of the membrane [1].

The cytoplasmic membrane is vital to the survival of all cells. Any agent that disrupts the stabilizing attractive forces between the structural components will compromise the integrity and function of the cytoplasmic membrane. Only two major groups of antibiotics used for therapy of microbial infections are active against the completed membrane, the polymyxins and the polyenes. Toxicity problems are associated with the use of all such drugs, due to their damaging effects upon mammalian membranes. The tissues of the kidneys, liver and nervous system are particularly susceptible to damage. This rather limited therapeutic value of membrane-active agents might change in the future, with increasing interest being paid to the peptide lantibiotics produced by Gram-positive bacteria (see Section 6.6.1) and the magainin peptides found in the skin of amphibians (see Section 6.6.2).

6.2 Detection of membrane damage

6.2.1 *The cytoplasmic membrane*

The first observations relating antimicrobial activity to membrane disruption [2] showed that low-molecular-weight cytoplasmic constituents are rapidly released from cells of *S. aureus* exposed to tyrocidin, an early cyclic peptide antibiotic [3]. A quantitative relationship between tyrocidin concentration, leakage and cell death was soon established. Measurement of the leakage of cytoplasmic constituents from drug-treated cells is a standard method of detecting membrane damage. Diverse chemical, spectrophotometric and potentiometric detection systems have been employed to measure the release of inorganic cations, amino acids, nucleosides and nucleotides from the metabolic pool. Selective ion electrodes (e.g. for potassium) have been particularly useful for the continuous *in situ* measurement of leakage without the need to remove the cells from suspension. Direct observation of treated cells by electron microscopy shows major disruption of the cell surface, the release of sections of outer membrane as small vesicles or blebs, and the coagulation of cytoplasm, but finer details of the effects upon the cytoplasmic membrane cannot be distinguished. The loss of integrity of the membrane can be monitored indirectly by measuring the activity of enzymes within it. For example, the respiratory activity of cell suspensions can be monitored continuously using an oxygen electrode. The aerobic respiratory activity (the rate of oxygen consumption in the presence of a substrate such as glucose) is very sensitive to the action of a membrane-active agent. In some cases, the response of the cells is very sensitive to the amount of membrane-active agent added. Low (non-lethal) concentrations can cause enhanced oxygen uptake, possibly due to a stimulation of membrane transport. Increasing the concentration to lethal levels causes progressive inhibition of respiration that correlates with the extent of membrane damage.

6.2.2 The outer membrane

Damage to the outer membrane of Gram-negative bacteria can be detected by studying the penetration of molecules that are normally excluded from the cells. For example, the fluorescent probes ANS (6-anilino-1-naphthalenesulfonic acid) and NPN (1-N-phenylnaphthylamine) fluoresce strongly under ultraviolet (UV) light when bound to the cytoplasmic membrane [4]. When whole cell suspensions of Gram-negative bacteria are exposed to the probes, the intact outer membrane prevents them from crossing the envelope. However, when the outer membrane is damaged by a membrane-active agent, the probes can penetrate to the underlying cytoplasmic membrane, causing fluorescence under UV irradiation. This sensitive method is now widely used to study outer membrane-permeabilizing agents. A fluorescent dansyl derivative of polymyxin B [5] that damaged the outer membrane also acted as a fluorescent probe when it penetrated to the cytoplasmic membrane. Other methods using whole cell suspensions rely on the release of periplasmic markers from cells with damaged outer membranes. Enzymes such as β-lactamases are useful, since their release can be easily detected with a chromogenic substrate such as nitrocefin. Information on the effects of agents that form pores in membranes can be derived from electrical conductivity measurements made on artificial membrane bilayers. Defined membrane potentials are applied across an artificial membrane dividing two electrode chambers. The membranes are effective insulators, and therefore the increased conductivity that is detected when agents are applied indicates the formation of pores in the membrane.

6.3 Membrane-damaging agents used as disinfectants, antiseptics and preservatives

The cytoplasmic membrane of microbial cells can be attacked by a wide range of chemicals, including phenols, alcohols and detergents (see Section 1.2). These membrane-active agents (*Table 6.1*) interfere with non-covalent interactions between the components of the membrane structure that maintain its integrity as a permeability barrier. Phenols denature membrane proteins, alcohols penetrate into the hydrophobic core of the membrane and cationic detergent molecules insert themselves between adjacent lipids and proteins. Loss of membrane integrity results in the release of essential cytoplasmic constituents and the cells rapidly die. Unfortunately, this very effective killing mechanism is not specific for microbial cells. Thus, while these agents can be used as disinfectants, antiseptics and preservatives, they are indiscriminate and so are not suitable for systemic treatment of microbial infections due to their disruption of mammalian cell membranes.

6.4 Antibacterial peptide antibiotics

In view of the poor selective toxicity of membrane-active agents, it may be surprising that some agents which act in this way can be used therapeutically to treat microbial infections. (*Figure 6.2*). However, there are a number of antibiotics which act in this way.

6.4.1 The tyrocidins

The tyrocidins are components of the antibiotic tyrothricin, which is produced by the Gram-positive soil bacterium, *Bacillus brevis*. They are peptides that are potent in disrupting the cytoplasmic membrane

Table 6.1: Some membrane-active agents used as disinfectants, antiseptics or preservatives

Phenolics:	cresols, xylenols, chlorocresol (4-chloro-3-methylphenol), chloroxylenol (4-chloro-3,5-dimethylphenol)
Alcohols:	ethanol, *iso*-propanol, 2-phenylethanol, 2-phenoxyethanol, chlorbutol (trichloro-*tert*-butanol), bronopol (2-bromo-2-nitropropan-1,3-diol)
Detergents:	quaternary ammonium compounds (e.g. benzalkonium chloride, cetyltrimethylammonium bromide, cetylpyridinium chloride)
Biguanides:	chlorhexidine, alexidine, polyhexamethylene biguanides

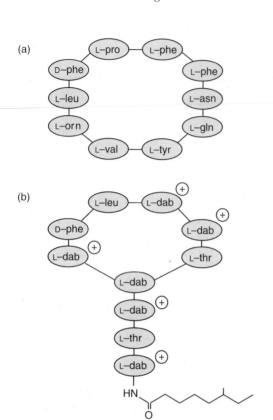

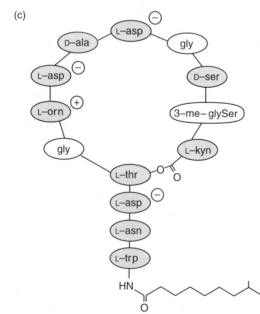

Figure 6.2: The structure of some peptide antibiotics: (a) tyrocidin A; (b) polymyxin B; (c) daptomycin.

of Gram-positive bacteria. Although they are too toxic for systemic therapy of infections, they are used in antiseptic throat lozenges. The tyrocidins are cyclic decapeptides which display marked surface activity and form molecular aggregates (micelles) in solution. The cyclic structure is important for their antimicrobial action because a linear analog of tyrocidin A, containing the same sequence of 10 amino acids, does not have antimicrobial activity.

6.4.2 *The polymyxins*

The polymyxins are a family of cyclic lipopeptide antibiotics produced by *Bacillus polymyxa*. Polymyxin B and polymyxin E (colistin) are used for treatment of serious Gram-negative bacterial infections, particularly those involving *P. aeruginosa* [6]. Their structure is related to that of the tyrocidins, comprising a cyclic peptide linked via a tripeptide to a branched-chain fatty acid residue. Like the tyrocidins, they contain the unusual D-stereoisomer of phenylalanine (only the L-isomer occurs in proteins). The polymyxins also contain positively charged diaminobutyric acid (DBA), which is not found in proteins. The combination of the positively charged polar peptides and the hydrophobic hydrocarbon tail gives the polymyxins properties similar to those of cationic detergents. The polymyxins damage both the outer and cytoplasmic membranes of Gram-negative bacteria. Lipopolysaccharide molecules located on the outer face of the outer membrane form the initial target of their interaction with the cells. The binding of polymyxins to the lipopolysaccharide destroys the permeability barrier of the outer membrane. Electron micrographs of polymyxin-treated cells show the outer membrane to be distorted. In extreme cases, segments of the outer membrane form blebs that protrude from the cell surface and become detached from the cell. This gross physical disruption may result from expansion of the outer face of the outer membrane where the polymyxin binds to the lipopolysaccharide. Once the outer membrane barrier has been destroyed, the polymyxin can penetrate through the cell wall to the cytoplasmic membrane, where it interacts with the phospholipids in the membrane bilayer. The positively charged DAB residues of the polymyxin bind to the phosphate residues in the polar head groups on the membrane surface, while the hydrophobic alkyl chains of polymyxin are inserted between the fatty acids. Destabilization of the phospholipid bilayer causes immediate leakage of cytoplasmic material. The hydrophobic alkyl chain of polymyxin is essential for the interaction with the cytoplasmic membrane, but not for the damaging effect upon the outer membrane.

This has been demonstrated with polymyxin B nonapeptide, prepared from polymyxin B by removal of the terminal fatty acid and the DAB residue with the proteolytic enzyme, ficin. Polymyxin B nonapeptide causes the same damage to the outer membrane as entire polymyxin B, but causes no damage to the cytoplasmic membrane and is not lethal towards bacteria. This selective outer membrane permeabilization may be exploitable, and attempts are being made to use agents such as polymyxin B nonapeptide to improve the penetration of other antimicrobial agents that are normally excluded by the outer membrane. Resistance to polymyxin is conferred by a plasmid-borne novel protein that has some homology with a viral acid protease [7].

Ranalexin from the skin of the bullfrog is a 20-residue peptide that resembles polymyxin in having a heptapeptide ring and exemplifies the animal products that may become useful in antimicrobial therapy [8].

6.4.3 Daptomycin

Daptomycin is a lipopeptide antibiotic that resembles the polymyxins, but the presence of aspartate residues gives it an overall negative charge. Its spectrum of antimicrobial activity and mechanism of action are quite different from those of polymyxin. Daptomycin is active against Gram-positive bacteria but not against Gram-negative bacteria. It has useful activity against strains of *S. aureus* which are resistant to most other groups of antibiotics. Originally isolated as resistant to the penicillin methicillin, these MRSA strains pose a major threat, particularly in hospital environments where they can be responsible for cross-infections. It is not clear how daptomycin works. The cytoplasmic membrane appears to be its initial target, but gross damage and leakage of cytoplasm does not ensue. The biosynthesis and assembly of the peptidoglycan, teichoic acid and lipoteichoic acid components of Gram-positive walls and membranes appear to be affected (see Section 3.2) possibly as a result of interference with export across the cell membrane.

Resistance to the glycopeptide antibiotics, especially vancomycin, is an emerging problem in the enterococci and MRSA against which it acts as a major life-saving antibiotic, but fortunately does not also confer resistance to the lipopeptides such as daptomycin [9].

6.4.4 Bacitracin.

Like daptomycin, this cyclic peptide antibiotic exerts a specific effect on the transfer of peptidoglycan precursors across the cytoplasmic membrane of bacteria by a lipid carrier, undecaprenyl phosphate (see Section 3.3.1.). This carrier is recycled from pyrophosphate to monophosphate, and the phosphatase activity responsible for this conversion is inhibited by bacitracin. The drug has been used for sterilization of the gut before surgery, but not for systemic therapy of infections because of its toxic side-effects.

6.5 Antifungal polyene antibiotics

The polyene group of antibiotics consists of nonpeptide antifungal compounds produced by *Streptomyces* species. Their structure is characterized by a large ring system, one side of which contains a rigid hydrophobic alkyl chain with a series of conjugated double bonds (the polyene region). The other side of the ring is a flexible saturated hydrophilic region containing a number of hydroxyl and keto groups. Amphotericin B (Fungilin) and nystatin are the most important members of the polyene group. Although amphotericin B causes extensive kidney and neurological damage ('Ampho the terrible' [10]), it is the drug of choice for treatment of life-threatening systemic fungal infections. Nystatin is toxic and is not given by injection or infusion for systemic use. It is used topically for fungal infections (e.g. vaginal candidiasis), but may also be given as oral drops and lozenges for the corresponding throat infection (thrush), because it is not absorbed from the gut (*Figure 6.3*).

The polyenes work in a different way to the polymyxins. They have a high affinity for ergosterol in fungal membranes that derives from hydrophobic interactions between the polyene region of the antibiotic and the fused ring system of the sterol. The polyene region of the drug binds to sterols located between phospholipids in the cytoplasmic membrane, and the entire drug molecule, including the hydroxylated

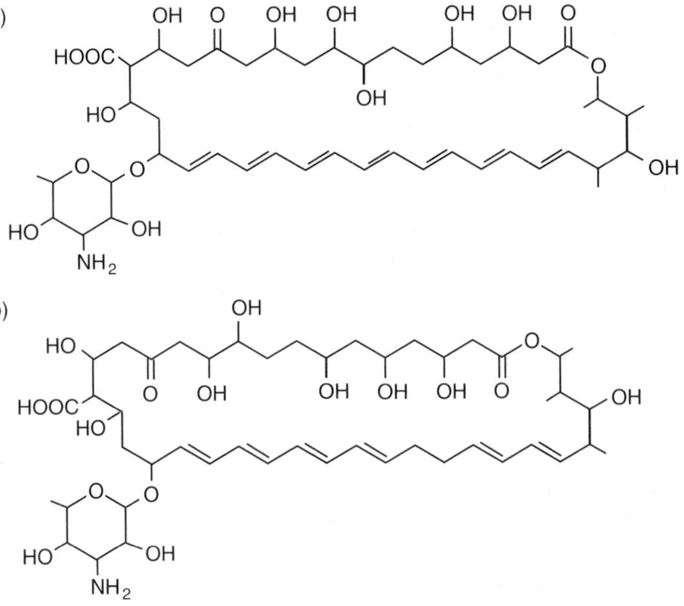

Figure 6.3: Structures of the polyene antifungal antibiotics: (a) amphotericin B; (b) nystatin.

hydrophilic side of the ring, is pulled into the membrane bilayer. A number of drug molecules associate together to generate hydrophilic pores in the membrane through which cytoplasmic components are released. Initially, only smaller cytoplasmic constituents such as potassium ions leak from the cells, suggesting that the pores formed initially are small, but larger molecules, including amino acids and nucleotides are released later. Loss of potassium leads to internal acidification, precipitation of the cytoplasm and cell death.

The polyenes have no antibacterial action because bacterial membranes do not contain sterols. However, their acute toxicity results from binding to cholesterol in mammalian cell membranes. The slight selectivity that they display towards fungi is due to their stronger affinity for ergosterol than for cholesterol. Because there are no other antifungal agents that match the polyenes in lethal antifungal activity, attempts have been made to reduce their toxicity to man. One approach has been to administer amphotericin B in the form of phospholipid liposomes, and this type of formulation is slightly less toxic to mammalian cells, but retains its antifungal activity.

6.6 Other peptides with the potential for antimicrobial use

6.6.1 Bacteriocins of Gram-positive bacteria

Bacteriocins are proteins produced by bacteria that kill related bacteria [11]. The classical bacteriocins produced by Gram-negative bacteria, such as the colicins of *E. coli*, have very selective effects upon their target bacteria. They enter the cells via surface receptors (e.g. outer membrane proteins), and then specifically inhibit cellular functions within the cells. Gram-positive bacteria also produce bacteriocins. Many are heat-stable peptides containing 30 to 60 amino acids, some of which show considerable promise as food preservatives and also for the treatment of infectious diseases. Unlike the polymyxin and tyrocidin peptide antibiotics, which are synthesized inside producer strains by a series of reactions involving multienzyme complexes [12], the bacteriocins are assembled at the ribosome. Most are translated as inactive prepeptides containing an N-terminal leader sequence and a C-terminal propeptide. The leader sequence is removed and, in some cases, amino acids in the C-terminal portion undergo chemical modifications to yield the active antimicrobial peptide. Post-translational amino acid modifications involve removal of a water molecule from serine and threonine residues, and formation of thioether rings between amino acid residues. The novel amino acids produced are didehydroalanine and didehydrobutyrine and the thioether amino acids lanthionine and β-methyl lanthionine. Such peptide bacteriocins containing lanthionine are referred to as lantibiotics and are produced by a range of Gram-positive bacteria (*Table 6.2*). The oral bacterium *Streptococcus salivarius* forms salivaricin A, a 22-residue peptide containing one lanthionine and two β-methyl lanthionine residues [13]. Mersacidin from *Bacillus subtilis* resembles the type B lantibiotics [14]. It is to be hoped that there may be many of these peptides yet to be discovered, and that synthetic variants may become available. The lantibiotic nisin, which is produced by the lactic acid bacterium *Lactococcus lactis*, was first described in 1928 and has a broad spectrum of activity against Gram-positive bacteria. It occurs naturally in some dairy products during processing, and it appears to be safe when ingested.

Lantibiotics form pores in the cytoplasmic membrane, resulting in leakage of metal ions and low-molecular-weight metabolites and the collapse of the electrochemical proton gradient across the membrane. Electrostatic interactions between positively charged lysine residues in the lantibiotic and the negatively charged phosphate head groups of the phospholipids are responsible for binding of the lantibiotics to the membrane surface. The electrochemical proton gradient across the membrane is needed to form the pores. It is thought that the lantibiotics form clusters of helices which span the membrane. The properties of the pores depend upon the lantibiotic, the composition of the phospholipids in the membrane and the membrane potential. The pores are unstable, lasting from a few milliseconds up to 30 sec. Gram-negative bacteria are generally not sensitive to lantibiotics unless the outer membrane barrier is removed to allow access to the cytoplasmic membrane. There also seems to be an inhibitory action against spores, possibly through interaction of the didehydroamino acids of nisin with sulfhydryl groups on membrane proteins of the germinating spores.

There are many other related peptides that do not contain lanthionine produced by lactic acid bacteria. Lactococcins A and B are small hydrophobic cationic peptides formed by *L. lactis*. Unlike the lantibiotics, the lactococcins must bind to specific receptors on other lactococci in order to interact with their membranes and kill the cells. Several lactococcin molecules form aggregates with the receptors, producing barrel-like structures which form pores when inserted into the membrane.

6.6.2 *Other antimicrobial peptides*

Gene-encoded antimicrobial peptides have been isolated from a wide variety of sources other than bacteria. Defensins are small (~30 amino acid residues) cationic antimicrobial peptides produced by mammalian polymorphonuclear leukocytes, macrophages and epithelial cells. They

Table 6.2: Examples of naturally occurring membrane-active antimicrobial peptides

Group	Name	Source	Amino acids	Structural features
Microbial lantibiotics:	Nisin	*Lactococcus lactis*	34	Lanthionine/modified residues
	Epidermin	*Staphylococcus epidermidis*	21	Lanthionine/modified residues
	Pep5	*Staphylococcus epidermidis*	33	Lanthionine/modified residues
	Subtilin	*Bacillus subtilis*	32	Lanthionine/modified residues
Human defensins:	HNP 1–4	Human neutrophils	29–33	6 disulfide-linked cysteines
	HD 5,6	Human intestinal cells	*c.* 30	6 disulfide-linked cysteines
Insect peptides:	Cecropins	Moths (*Hyalophora cecropia*)	35–39	
		Flies (*Sarcophaga peregrina*)		
Amphibian skin peptides:	Magainins	Frogs (*Xenopus laevis*)	21–27	4 lysines, amphipathic
Arthropod peptides:	Tachyplesins	Horseshoe crab hemocytes	17–18	4 disulfide-linked cysteines
		(*Tachypleus* and *Limulus* spp.)		5 arginines
Plant seed peptides:	*Mj*-AMP1 and 2	Four o'clock plant (*Mirabilis jalapa* L.)	36,37	6 disulfide-linked cysteines
	Ac-AMP1 and 2	Amaranth (*Amaranthus caudatus*)	29,30	6 disulfide-linked cysteines

have a conserved framework structure involving six disulphide-linked cysteine residues, and they act on prokaryotic and eukaryotic cells by disrupting the cell membrane. Other antimicrobial defense peptides have been identified in protective secretions from the skin of amphibians (frogs and toads), from invertebrates (horseshoe crabs), insects (flies and moths) and plant seeds. The magainins show the greatest promise for therapy of infections. They were discovered in 1987 in skin secretions of *Xenopus laevis*, and display a broad spectrum of activity against Gram-positive and Gram-negative bacteria, fungi and protozoa. Many analogs have been made, and some are undergoing clinical trials for topical treatment of skin infections. Unlike many of the other membrane-active agents, the magainins do not appear to exert toxic effects upon mammalian cells. They do not lyse red blood cells or circulating lymphocytes, and may therefore be useful for systemic therapy of infections. The selective activity towards microbial cell membranes is probably due to the affinity of the magainins for the anionic phosphate groups of phospholipids (especially phophatidyl glycerol and cardiolipin) that are accessible for binding on the surface of microbial membranes. In solution, the magainins adopt random flexible conformations, but when bound to the membrane surface they form an α-helical structure, with cationic lysine residues on one side of the helix and hydrophobic residues on the other side. Whereas the lantibiotics penetrate the membrane, the magainins appear to remain on the membrane surface in aggregates. Disruption of the membrane structure leads to leakage of cellular contents and death of the cells.

Pseudomycins are novel lipopeptides from *Pseudomonas syringae* that contain unusual amino acids such as chlorthreonine that have proved fungicidal *in vitro* to pathogens like *Aspergillus fumigatus*, *Candida albicans* and *Cryptococcus neoformans*. They are also effective against the plant pathogens, such as *Bipolaris spicifera* and *Coccidioides immitis*, that are killing immunocompromised individuals [10].

6.7 Inhibitors of sterol synthesis: imidazoles and triazoles

The azole antifungal drugs (*Figure 6.4*) inhibit the synthesis of the sterol components of the fungal membrane by inhibiting a reaction that is dependent on cytochrome P_{450}, namely the 14α-demethylation of lanosterol [15], a precursor of ergosterol. The drugs cause the rapid development of defects in membrane permeability, with loss of cytoplasmic components leading to similar effects to the polyenes. The azoles can be given orally [16] to treat systemic fungal infections, as well as topically in dermatological fungal infections. Ketoconazole (Nizoral) may have serious effects on the liver, probably due to inhibition, but is nevertheless commonly used orally and topically. Miconazole (Daktarin,

Monistat) is not so frequently used. Because these drugs inhibit steroid metabolism to a degree, and reduce testosterone synthesis, one of their side-effects may involve the female development of male breast tissue (gynaecomastia). Oral and esophageal thrush may be treated with miconazole, ketoconazole, and clotrimazole (Canesten, Gyne-Lotremin, Mycelex-G), as well as the more recent triazole, fluconazole (Diflucan), which are often prescribed in ointment and pessaries for vaginal candidiasis. Fluconazole is also used to treat cryptococcal meningitis.

Resistance to azoles has emerged in *C. albicans* [17] and other fungi, and this may be associated with a reduced affinity of the drugs for mutant cytochrome P_{450} 14α-demethylase [18]. Sensitivity to azoles may also depend on the properties of accessory enzymes that are involved in electron transfer to the cytochrome P_{450}. In yeast, disruption of the NADPH reductase that services the demethylase increases

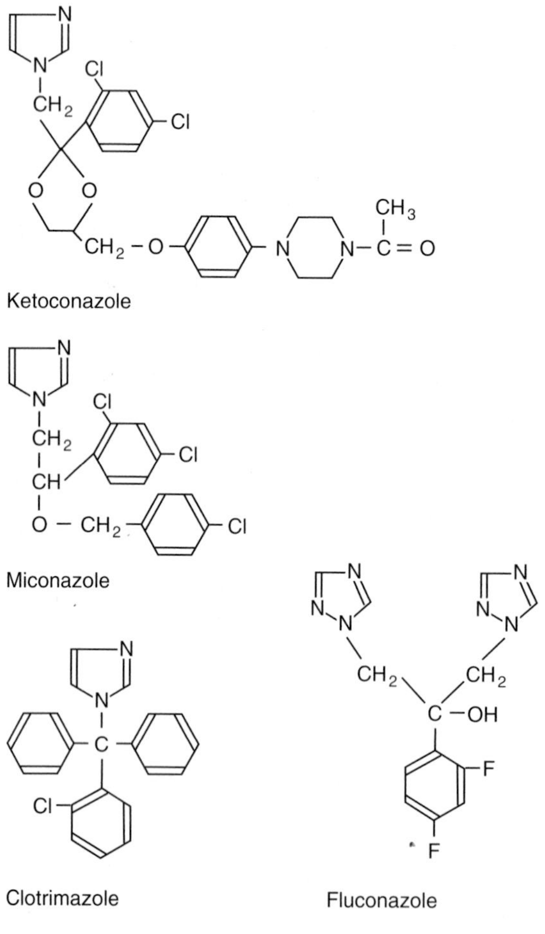

Figure 6.4: Azole antifungal drugs.

the sensitivity (i.e. decreases resistance) to ketoconazole by 200-fold [19]. This change can, in turn, be reversed by the provision of a cytochrome b_5, which presumably acts as an alternative source of electrons [20].

References

1. Benz, R. (1988). *Annu. Rev. Microbiol.*, **42**, 359–394.
2. Hotchkiss, R.D. (1944) *Adv. Enzymol.*, **4**, 153–199.
3. Gale, E.F. and Taylor, E.S. (1947) *J. Gen. Microbiol.*, **1**, 77–84.
4. Loh, B., Grant, C. and Hancock, R.E.W. (1984) *Antimicrob. Agents Chemother.*, **26**, 546–551.
5. Newton, B.A. (1956) *Bacteriol. Rev.*, **20**, 14–27.
6. Hancock, R.E.W. and Wong, P.G.W. (1984) *Antimicrob. Agents Chemother.*, **26**, 48–52.
7. Roland, K.L., Esther, C.R. and Spitznagel, J.K. (1994) *J. Bacteriol.*, **176**, 3589–3597.
8. Clark, D.P., Durell, S., Maloy, W.L. and Zasloff, M. (1994) *J. Biol. Chem.*, **269**, 10849–10855.
9. Derlot, E and Courvalin, P. (1991) *Am. J. Med.*, **91**, 82S–85S.
10. Potera, C. (1994) *Science*, **265**, 605.
11. Jack, R.W., Tagg, J.R. and Ray, B. (1995) *Microbiol. Rev.*, **59**, 171–200.
12. Kleindorf, H. and von Dohren, H. (1990) *Eur. J. Biochem.*, **192**, 1–15.
13. Ross, K.F., Ronson, C.W. and Tagg, J.R. (1993) *Appl. Environ. Microbiol.*, **59**, 2014–2021.
14. Bierbaum, G., Brotz, H., Koller, K.P. and Sahl, H.G. (1995) *FEMS Microbiol. Lett.*, **127**, 121–126.
15. Bodey, G.P. (1992) *Clin. Infect. Dis.*, **14** (Suppl. 1), S161–S169.
16. Hay, R.J. (1991) *J. Antimicrob. Chemother.*, **28** (Suppl. A), 35–46.
17. Ruhnke, M., Eigler, A., Tennagen, I., Geiseler, B., Engelman, E. and Trautmann, M. (1994) *J. Clin. Microbiol.*, **32**, 2092–2098.
18. van den Bossche, H., Marischal, P., Gorrens, J., Bellens, D. and Moereels, P.A.J. (1990) *Biochem. Soc. Trans.*, **18**, 56–59.
19. Sutter, T.R. and Loper, J.C. (1989) *Biochem. Biophys. Res. Commun.*, **160**, 1257–1266.
20. Truan, G., Epinat, J.-C., Rougeulle, C., Cullin, C. and Pompon, D. (1994) *Gene*, **149**, 123–127.

Chapter 7

Chemotherapy of protozoal diseases

7.1 Malaria

Among protozoal diseases malaria alone kills an estimated 1–2 million people each year [1]. The disease is caused by certain species of *Plasmodium* (*P. falciparum*, *P. malariae*, *P. ovale*, *P. vivax*) which parasitize erythrocytes (red blood cells). The parasite is transmitted from person to person via female anopheline mosquitoes. Most deaths result from severe falciparum malaria, a disease characterized by the binding of parasitized erythrocytes to the inner surfaces of small blood vessels [2]. In some cases, susceptibility to severe disease has been found to correlate with a variant form of the TNF-α gene promoter [3].

Chemotherapy is aimed at certain stages in the parasite's life cycle (*Figure 7.1*). Drugs effective at a given stage may have little or no activity at other stages (*Table 7.1*). Some antimalarial drugs are useful for both treatment and prophylaxis.

7.1.1 Antifolate drugs: proguanil (Paludrine) and pyrimethamine–sulfadoxine (Fansidar)

These drugs inhibit the production of tetrahydrofolate (THF) in sensitive strains of the parasite. THF deficiency kills the parasite because THF is essential for biosynthesis (see Section 2.4).

In combination with chloroquine (oral, weekly) or another antifolate drug, *proguanil* (oral, daily) is one of several options for antimalarial prophylaxis. The drug has a good record of safety and tolerance. Proguanil is oxidized in the liver to the active form, cycloguanil (*Figure 7.2*). Within the parasite, cycloguanil binds competitively to dihydrofolate reductase (DHFR; see *Figures 2.3* and *2.6*), thus inhibiting THF synthesis. In sensitive strains, this drug kills tissue schizonts as well as blood schizonts (although not hypnozoites). Interestingly, the efficiency

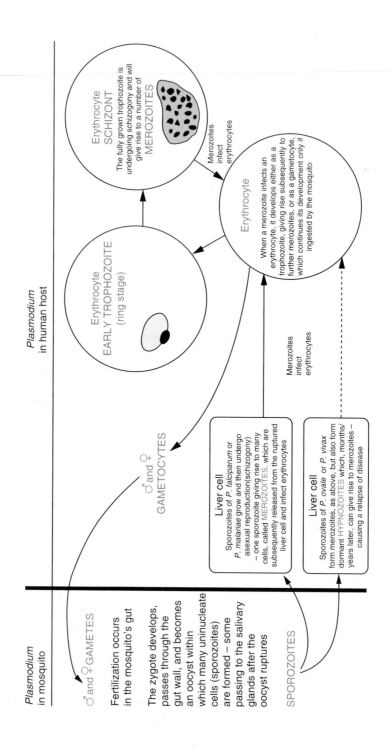

Erythrocyte
SCHIZONT

The fully grown trophozoite is undergoing schizogony and will give rise to a number of MEROZOITES

Merozoites infect erythrocytes

Erythrocyte

When a merozoite infects an erythrocyte, it develops either as a trophozoite, giving rise subsequently to further merozoites, or as a gametocyte, which continues its development only if ingested by the mosquito

Erythrocyte
EARLY TROPHOZOITE
(ring stage)

Merozoites infect erythrocytes

♂ and ♀
GAMETOCYTES

Liver cell

Sporozoites of *P. falciparum* or *P. malariae* grow and then undergo asexual reproduction(schizogony) – one sporozoite giving rise to many cells, called MEROZOITES, which are subsequently released from the ruptured liver cell and infect erythrocytes

Liver cell

Sporozoites of *P. ovale* or *P. vivax* form merozoites, as above, but also form dormant HYPNOZOITES which, months/ years later, can give rise to merozoites – causing a relapse of disease

Plasmodium
in human host

Plasmodium
in mosquito

♂ and ♀ GAMETES

Fertilization occurs in the mosquito's gut

The zygote develops, passes through the gut wall, and becomes an oocyst within which many uninucleate cells (sporozoites) are formed – some passing to the salivary glands after the oocyst ruptures

SPOROZOITES

Table 7.1: Antimalarial drugs

Drug	Chemical description	Active against[a]
Artemisinin	Sesquiterpene lactone endoperoxide	BS
Atovaquone[b]	Hydroxynaphthoquinone	BS
Chloroquine	4-Aminoquinoline	BS
Halofantrine	9-Phenanthrene methanol	RS?
Mefloquine	4-Aminoquinoline	BS
Primaquine	8-Aminoquinoline	TS, HYP
Proguanil	Diguanide	TS, BS
Pyrimethamine	Diaminopyrimidine	BS
Quinidine	Quinoline-type (alkaloid)	BS
Quinine	Quinoline-type (alkaloid)	BS
Sulfadoxine	Sulfonamide	BS
WR238605[b]	8-Aminoquinoline	HYP, GAM

[a]For stages in the plasmodial life cycle see Figure 7.1. BS = blood schizont; TS = tissue schizont; HYP = hypnozoite; GAM = gametocyte; RS = ring stage.
[b]Candidate antimalarial currently being evaluated.

with which proguanil is converted to cycloguanil differs in different human populations, so that the usefulness of the drug for some groups has been questioned [4].

Resistance to cycloguanil can involve a mutant form of DHFR in which threonine replaces serine at residue 108 and an alanine-to-valine change occurs at residue 16 [5]. These changes presumably decrease the affinity of cycloguanil for the target enzyme.

The oral combination *pyrimethamine–sulfadoxine* is used in some areas for the treatment of clinically uncomplicated, chloroquine-

---

Figure 7.1: The life cycle of the malaria parasite (*Plasmodium*) (diagrammatic) showing the stages which are targets for antimalarial drugs (*Table 7.1*). Sporozoites, injected into the bloodstream via mosquito bite, penetrate liver cells. A sporozoite grows and develops into a *tissue schizont* which undergoes asexual reproduction (*schizogony*) to produce a number of cells called *merozoites*. On rupture of the liver cell, merozoites enter the bloodstream and infect erythrocytes. Within an erythrocyte, a merozoite develops as a *trophozoite* (growing stage), a young trophozoite being known as the *ring stage* owing to its appearance in stained preparations under the microscope. The mature trophozoite (*blood schizont*) undergoes schizogony, many merozoites being released when the erythrocyte ruptures. These merozoites infect fresh erythrocytes. Some merozoites become *gametocytes* (i.e. cells which have the potential to develop into male and female gametes). If ingested by the mosquito, gametocytes give rise to gametes and the sexual phase of the life cycle follows.

In two species of *Plasmodium*, the sporozoite can form a dormant stage (*hypnozoite*) in liver cells. For this form of malaria, a 'radical cure' involves treatment with primaquine (to eradicate hypnozoites) as well as other drug(s) to deal with intra-erythrocytic parasites.

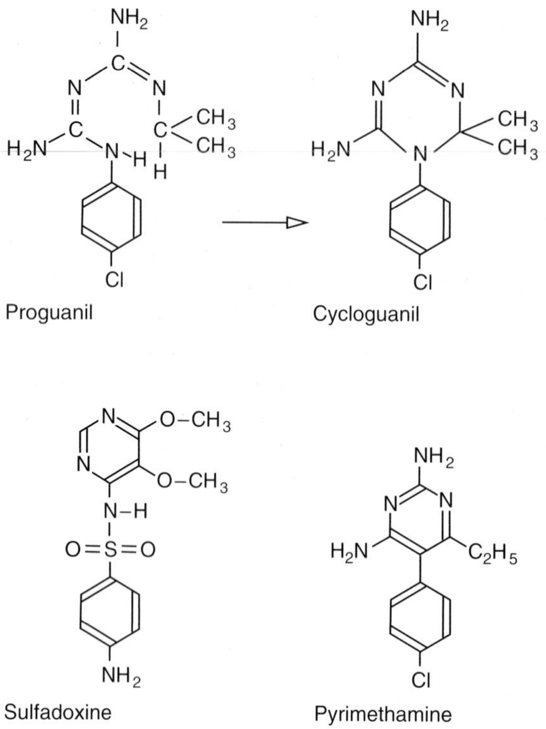

Proguanil

Cycloguanil

Sulfadoxine

Pyrimethamine

Figure 7.2: Some antifolate drugs. Proguanil is modified *in vivo* to the active form of the drug, cycloguanil. Sulfadoxine and pyrimethamine synergistically affect tetrahydrofolate (THF) formation (see also *Figure 2.4*).

resistant falciparum malaria [6]. Both drugs are absorbed rapidly and eliminated slowly, typically allowing single-dose therapy. Contra-indications include early pregnancy owing to the possible teratogenicity of pyrimethamine. The possibility of severe skin reactions [7] tends to preclude prophylactic use.

Sulfadoxine and pyrimethamine synergistically block the synthesis of dihydrofolate (DHF) and its reduction to THF, respectively (see *Figure 2.4*). Observations *in vitro* and *ex vivo* indicate that the late trophozoite stage is affected.

In *Plasmodium*, resistance to pyrimethamine can arise in at least two ways. First, amplification of the DHFR gene [8], leading to over-production of DHFR and consequent relief from the inhibitory action of the drug. Second, point mutations [9] that alter the amino acid sequence of DHFR and decrease the affinity of binding of pyrimeth-amine. Amplification of the DHFR gene can be selected for in the laboratory, but appears not to occur in field isolates of resistant *P. falci-parum*; rather, such isolates seem typically to have variant forms of DHFR [10].

In pyrimethamine-resistant isolates, DHFR contains (as an essential, and minimum, change) asparagine in place of serine at residue 108. Higher levels of resistance correlate with further specific mutations in the DHFR gene. Wild-type and mutant DHFR genes from *P. falciparum* have been expressed in bacteria, and the mutant DHFRs formed in this way have been shown to exhibit a lower affinity for pyrimethamine [11].

Simultaneous resistance to pyrimethamine and proguanil can occur [12] but seems to require at least two or three appropriate mutations. The spread of cross-resistance might be discouraged by using the drugs in combination, but the parasite population is likely to contain a pool of mutant DHFR genes which would tend to undermine such a strategy.

Sulfadoxine-resistant strains of *P. falciparum*, with differing levels of resistance, show sequence variation in the dihydropteroate synthetase (DHPS, see Section 2.4) gene [13], suggesting that resistance to this drug may involve the effects of mutation on the target enzyme.

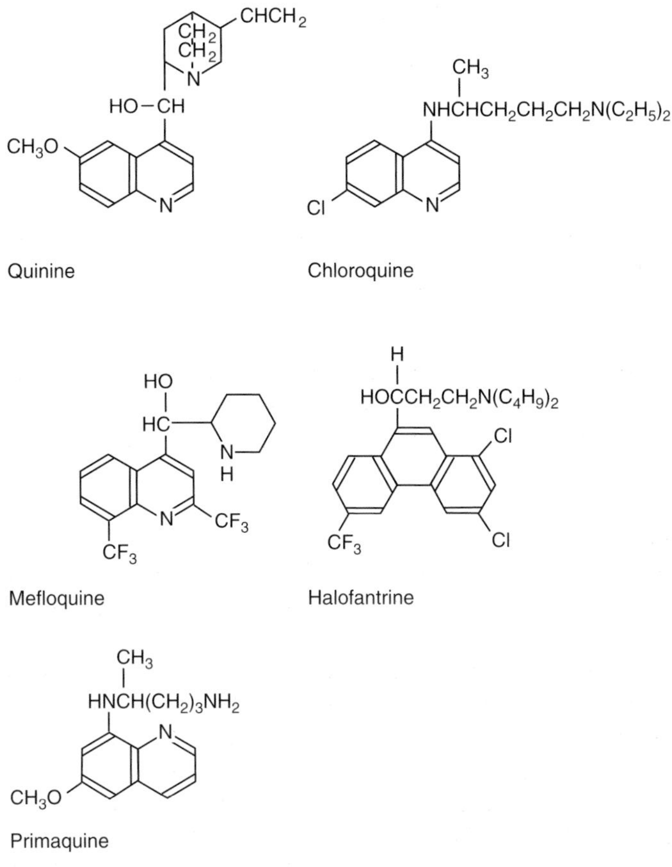

Figure 7.3: Quinine, chloroquine, mefloquine, primaquine and halofantrine.

7.1.2 Quinoline-type antimalarials: quinine, chloroquine and mefloquine

These drugs (*Figure 7.3*) lyse sensitive parasites within erythrocytes. Their mechanism of action is still not fully understood, but they appear to disrupt the parasite's normal detoxification of ferriprotoporphyrin IX (FP), a by-product formed during digestion of the erythrocyte's hemoglobin. Free FP, which can lyse the parasite, is normally detoxified by polymerization within the parasite's food vacuole (*Figure 7.4*). Polymerization of FP appears to involve 'heme polymerase' activity [14],

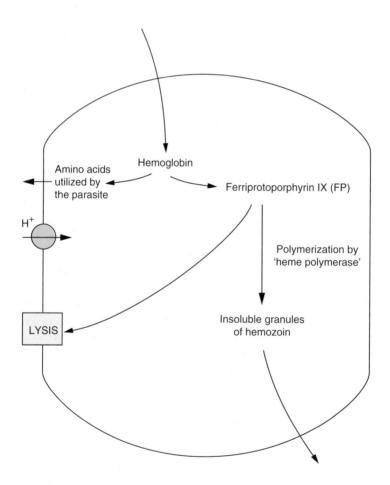

Figure 7.4: Digestion of hemoglobin in the food vacuole of *Plasmodium*. Acidity within the vacuole results from the inward pumping of protons. Ferriprotoporphyrin IX (FP) is normally detoxified by polymerization to hemozoin. Chloroquine and related antimalarials appear to inhibit polymerization, allowing free FP to bring about lysis.

and quinoline-type drugs possibly interact with, and inhibit, the heme polymerase, thus allowing free FP to lyse the cell.

The mode of action of chloroquine. The mechanism of action has been studied by using a chloroquine analog which, when photoactivated, labels those proteins which bind to it within parasitized erythrocytes [15]. Such photoaffinity labeling has identified two proteins (42 kDa and 33 kDa). Labeling by the analog was competitively inhibited by chloroquine (but not e.g. by (structurally-related) salicylamide), indicating that the labeled proteins are chloroquine-binding proteins. These proteins possibly correspond to 'heme polymerase'. Competitive inhibition of the photoaffinity labeling was also shown by quinine, perhaps indicating that the labeled proteins are targets common to the quinoline-type anti-malarials.

Resistance to quinoline-type drugs. Most studies have been carried out on chloroquine (the drug most widely used for uncomplicated malaria). Despite much research, the mechanism of resistance is still unknown.

Resistant and sensitive strains of the parasite differ consistently in their ability to accumulate chloroquine intracellularly, lower levels of the drug being found in resistant strains. Thus, phenotypic resistance seems to require that intracellular target structure(s) be exposed to lower concentrations of the drug. Studies have therefore focused on ways in which chloroquine enters and leaves the parasite (and its food vacuole) as such movements of the drug determine intracellular levels.

Chloroquine is a weak base, and its movement across membranes is influenced by its state of ionization. The nonprotonated form can diffuse freely across membranes. However, within the (acidic) food vacuole chloroquine is protonated so that it cannot diffuse out via the vacuolar membrane. An appropriate level of acidity in the vacuole therefore encourages accumulation of the drug at its putative site of action.

Current ideas on the mechanism(s) of resistance are outlined below.

Within the parasite's food vacuole, acidity is regulated by a proton pump located in the vacuolar membrane. Experiments with a proton-pump inhibitor have suggested that resistance to chloroquine might involve weakened pump activity leading to lower vacuolar acidity and (hence) to lower levels of the accumulated drug [16]. However, this explanation may be inadequate since the implied mode of drug accumulation (inward diffusion and protonation) may not, by itself, be sufficient to account for the level of uptake in sensitive strains [17].

In *P. falciparum*, the vacuolar membrane contains a 'permease' (transport system) [18] encoded by the gene *pfmdr1*. The permease (designated Pgh1) is a member of a superfamily of so-called ATP-binding cassette (ABC) transporters which occur in organisms as diverse as mammals and bacteria and which mediate ATP-dependent transmembrane transport of various substances. Recently [19], *pfmdr1* was expressed in Chinese hamster ovary (CHO) cells and caused an increased uptake of chloroquine, the permease apparently localizing in the membrane of the lysosome and mediating transport into the organelle. This has suggested that Pgh1 might be involved e.g. in the uptake of chloroquine by the food vacuole in *P. falciparum*. Consistent with this suggestion is the finding that experimental selection for chloroquine resistance (by chloroquine-limited growth) causes 'deamplification' of *pfmdr1* and decreased amounts of Pgh1 concomitant with increased resistance to chloroquine [20].

A different mechanism of resistance was suggested by the apparent vigorous expulsion of chloroquine from resistant parasites by an energy-dependent process [21]. Such efflux is reminiscent of the mechanism of resistance in multidrug-resistant mammalian tumor cells. In these cells, an overexpressed ABC transporter, the so-called P-glycoprotein (encoded by the multidrug-resistance gene *mdr*), gives rise to the multidrug-resistance (MDR) phenotype by pumping out therapeutic drugs. In tumor cells the MDR phenotype can be reversed e.g. by calcium-channel blockers such as verapamil. In *P. falciparum*, resistance to chloroquine can be reversed by verapamil (and by certain other compounds), suggesting (by analogy) that such compounds may inhibit an efflux mechanism in chloroquine-resistant *P. falciparum*. Conceivably, reversal of resistance has therapeutic possibilities, although so far there have been no reports of reversal being used therapeutically in man.

Recently, chloroquine-resistant isolates of *P. falciparum* were selected for mefloquine resistance (by mefloquine-limited growth) to determine the effect on *pfmdr1* expression and chloroquine resistance [22]. The resulting mefloquine-resistant isolates overexpressed Pgh1 and were less resistant to chloroquine – though more resistant to quinine (and to halo-fantrine). Why is the increased resistance to mefloquine linked to a decrease in chloroquine resistance? This inverse relationship, which had been noted previously [20], is not understood. The observed cross-resistance between quinine and mefloquine has suggested that resistance to mefloquine in areas where the drug has *not* been used therapeuti-cally may have arisen through exposure to quinine [22].

Our view of chloroquine resistance has been further complicated by experiments [23] which link resistance with a site on chromosome 7, *pfmdr1* being on chromosome 5. Resistance may therefore involve two or more genes on different chromosomes.

Quinine in chemotherapy. Parenteral quinine is currently the drug of choice for treatment of severe malaria owing to the low incidence of strains resistant to this form of therapy [6]. The drug (*Figure 7.3*) is about 80% protein-bound in healthy individuals but this rises to 90% or more in patients with malaria owing to elevated levels of the main binding protein, α_1-acid glycoprotein. Only the unbound drug is useful therapeutically. Hypersensitivity to quinine occurs rarely. The diastereo-isomer of quinine, *quinidine*, is used instead of quinine in the USA; it binds less to protein but is more cardiotoxic.

Chloroquine in chemotherapy. This inexpensive synthetic drug (*Figure 7.3*) is used in oral form in some malaria-endemic countries for the treat-ment of uncomplicated falciparum malaria in the indigenous population, being more effective than quinine against sensitive strains. It is also used, parenterally, to treat severe malaria although, owing to widespread resistance, alternative drugs are generally recommended for the severe disease.

Chloroquine is the drug of choice for malaria caused by *P. malariae*, *P. ovale* or *P. vivax* since most strains of these species are sensitive to the drug (resistant strains have been reported e.g. from Papua New Guinea). This type of malaria (which is rarely life-threatening) usually responds to other quinoline-type drugs, but antifolates may be less effective.

Chloroquine is also used for prophylaxis, e.g. in combination with proguanil.

Mefloquine in chemotherapy. This (oral) drug (*Figure 7.3*) is well absorbed, extensively protein-bound, and excreted in urine and bile with an elimination half-life of approximately 15–30 days. The drug is used for prophylaxis (e.g. in Africa, where chloroquine/proguanil is now often ineffective); for the treatment of uncomplicated falciparum malaria caused by strains resistant to chloroquine and antifolates; and (e.g. via nasogastric tube) for the treatment of severe malaria. Although generally well tolerated, side-effects may include nausea, vom-iting, diarrhea and dizziness. Neuropsychiatric effects may be important [24]. Prophylactic use is absolutely contraindicated in early pregnancy.

7.1.3 Halofantrine

This (oral) drug (*Figure 7.3*) is used for uncomplicated malaria caused by strains that are resistant to other drugs. Unlike quinine and the anti-folates, halofantrine seems to be active against the ring-stage trophozoite (*Figure 7.1*) [25]. The mechanism of action is unknown. Poor and variable absorption [26] preclude prophylactic use. Rarely, cardiac arrhythmias are associated with high levels of the drug [27].

7.1.4 Primaquine and WR238605

Primaquine is a quinoline-type drug (*Figure 7.3*) but, unlike the others, it is active against the hypnozoite stage (*Figure 7.1*) of *P. ovale* and *P. vivax*, and is used to avoid relapse. The drug is rapidly absorbed from the gut and extensively metabolized to pharmacologically active products, but the mode of action is unknown. Certain adverse effects (e.g. hemolysis, methemoglobinemia) are especially likely in patients who are deficient in glucose 6-phosphate dehydrogenase (G6PD).

WR238605 is a primaquine analog, being developed at the Walter Reed Army Institute of Research (US Army), which has yielded promising results in animal systems [28]. The drug is active against hypnozoites and gametocytes, and it may be able to replace primaquine.

7.1.5 Other antimalarials

Artemisinin (*qinghaosu*) is an orally administered sesquiterpene lactone endoperoxide (*Figure 7.5*), obtained from the traditional Chinese medicinal herb qinghao (*Artemisia annua*) [29], which is a rapidly acting blood schizonticide. Apparently, iron within the parasite activates the compound, producing potent free radicals [30]. The parenteral derivatives *artemether* and *artesunate* have been used successfully (e.g. in Indochina) for the treatment of severe malaria [31]. Some derivatives have shown neurotoxicity in animal studies [32].

Atovaquone is a hydroxynaphthoquinone that is being evaluated as a potential oral antimalarial. It inhibits plasmodial electron transport, apparently by interacting with cytochrome b in the bc_1 complex [33, 34]. Such inhibition blocks electron transfer from the enzyme system dihydroorotate dehydrogenase, thus blocking a vital oxidative step in the biosynthesis of pyrimidines (essential components of nucleic acids).

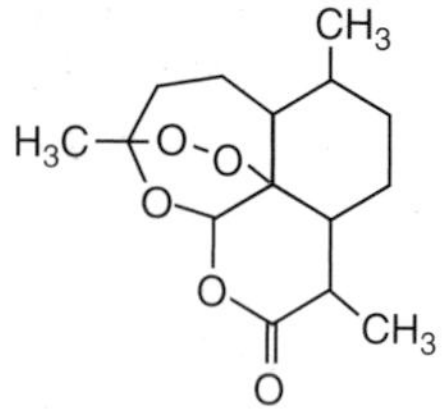

Figure 7.5: Artemisinin.

7.2 Cryptosporidiosis

Cryptosporidiosis in the immunocompetent typically involves profuse watery diarrhea which resolves in 1–2 weeks. In the severely immunocompromized it is a life-threatening disease. No effective drugs are available [35].

7.3 Toxoplasmosis

In immunocompromized patients, acquired toxoplasmosis commonly presents as toxoplasmic encephalitis, a condition invariably fatal if untreated. Treatment usually involves an antifolate combination, often pyrimethamine and sulfadiazine, although patients intolerant of sulfonamides may be treated satisfactorily with a combination of pyrimethamine and the lincosamide antibiotic *clindamycin* [36]. Clindamycin binds to the 50S ribosomal subunit, presumably blocking protein synthesis (Section 5.10) in the parasite's mitochondria.

Clindamycin is also used for prophylaxis in the newborn to prevent retinochoroiditis arising through congenital infection.

7.4 African trypanosomiasis (sleeping sickness)

Several antitrypanosomal drugs inhibit the synthesis of polyamines – compounds which have various functions (e.g. in protein synthesis) and which are essential for cell viability.

Pentamidine, used for over 50 years, is a substituted dibenzamidine which apparently inhibits the trypanosomal enzyme *S*-adenosyl-L-methionine decarboxylase (*Figure 7.6*). Hypotension is a common side-effect.

Eflornithine (DL-α-difluoromethylornithine; DFMO) inhibits ornithine decarboxylase (ODC) (*Figure 7.6*), and is effective against *Trypanosoma brucei gambiense* (although not against the alternative causal agent, *T. brucei rhodesiense*). The therapeutic use of eflornithine may depend on a slow turnover of trypanosomal ODC compared with that of mammalian ODC. Such a differential appears to exist in the case of a cattle-infecting trypanosome [37]. Eflornithine readily enters the cerebrospinal fluid (CSF) and is therefore useful in late-stage trypanosomiasis (when the central nervous system (CNS) is involved).

Melarsoprol is a trivalent arsenical compound effective against both causal agents of the disease. The drug's specific target is trypanothione (N^1,N^8-bis[glutathionyl]spermidine) [38] (*Figure 7.6*), which interacts with the drug to give an adduct (Mel T) that inhibits the enzyme trypanothione disulfide reductase; this enzyme (not present in the mammalian host) plays a vital role in regulating the thiol/disulfide balance in the parasite. Compared with eflornithine, melarsoprol penetrates the CNS poorly but is nevertheless regarded as the most active trypanocidal drug available [39] and has been used for many years against the late-stage disease. An important side-effect, drug-induced encephalopathy, occurs in up to 10% of patients.

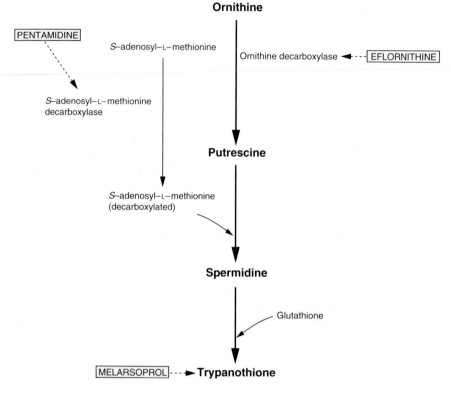

Figure 7.6: The sites of action of some antitrypanosomal drugs in the polyamine biosynthetic pathway. The reaction between putrescine and decarboxylated *S*-adenosyl-L-methionine yields methylthioadenosine (not shown) and spermidine.

Suramin, a sulfated naphthylamine, has been used for over 50 years but its mode of action is unknown, although it inhibits many enzymes. Owing to poor CNS penetration it is not effective in the late-stage disease.

7.5 Amebic dysentery

Acute amebic dysentery, caused by *Entamoeba histolytica*, has been treated e.g. with the alkaloid emetine (*Figure 7.7*). Emetine inhibits translocation in protein synthesis (Section 5.2.4), apparently by inter-acting with the 40S ribosomal subunit of *E. histolytica*. Altered 40S subunits have been found in emetine-resistant parasites. More commonly, nitroimidazoles, such as metronidazole (see Section 7.6), are used for the acute disease. Asymptomatic cyst-shedders in non-endemic countries are sometimes treated with diloxanide furoate.

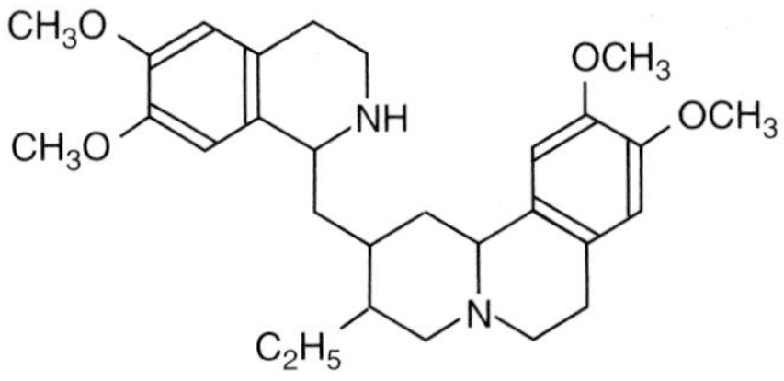

Figure 7.7: Emetine, an alkaloid drug used for the treatment of amebic dysentery. This compound is found in ipecacuanha root, an extract of which has been used for many years in South America for the treatment of this disease.

7.6 Trichomoniasis (vaginitis) and giardiasis

These diseases (caused by *Trichomonas vaginalis* and *Giardia intestinalis/ G. lamblia*, respectively) have been treated with drugs whose activity depends on their reduction within the parasite. The drugs include nitro-furans (e.g. furazolidone) and 5-nitroimidazoles (e.g. metronidazole, *Figure 7.8*). Reduction of the drugs gives rise to nitro anion radicals

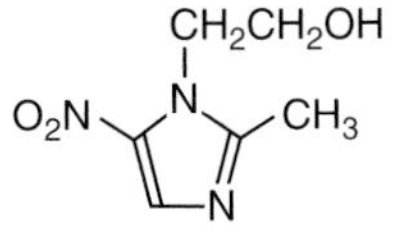

Figure 7.8: Metronidazole.

whose precise lethal target(s) within eukaryotic parasites are apparently unknown.

In *T. vaginalis*, the (ferredoxin-dependent) reduction of metronidazole occurs within the hydrogenosome, an organelle in which pyruvate is decarboxylated to acetyl phosphate, under anaerobic or near-anaerobic conditions, by pyruvate:ferredoxin oxidoreductase.

Ferredoxin is not essential for the reduction of nitrofurans [40].

Nitro anion radicals can reduce any oxygen present to form cytotoxic oxygen species (e.g. superoxide), a reaction which regenerates the drug from its radical. Only low steady-state levels of superoxide may be formed if the drug has a highly negative redox potential (e.g. metron-idazole, $E'_7 = -486$ mV), but significant levels may occur if the drug has a less negative potential, e.g. nitrofurantoin (a nitrofuran) ($E'_7 = -264$ mV). Thus, reactive oxygen species may be important in the mode of action of nitrofurantoin but not in that of metronidazole [40].

In *T. vaginalis*, resistance to metronidazole may involve defective oxygen scavenging, this allowing futile cycling of the drug and its anion with consequent lowering of the drug's efficacy [41]. In *G. intestinalis*, resistance has been associated with elevated levels of nicotinamide adenine dinucleotide phosphate (NADPH) oxidase, and it may be that the oxidized form of the coenzyme (NADP+) competes with metronidazole for the parasite's available reducing power [42].

References

1. World Health Organization (1993) *A Global Strategy for Malaria Control.* World Health Organization, Geneva.
2. Pasloske, B.L. and Howard, R.J. (1994) *Annu. Rev. Med.,* **45**, 283–295.
3. McGuire, W., Hill, A.V., Allsopp, C.E., Greenwood, B.M. and Kwiatkowski, D. (1994) *Nature,* **371**, 508–510.
4. Edstein, W., Shanks, G.D. and Teja-Isavadharm, P. (1994) *Br. J. Clin. Pharmacol.,* **37**, 67–70.
5. Foote, S.J., Galatis, D. and Cowman, A.F. (1990) *Proc. Natl Acad. Sci. USA,* **87**, 3014–3017.
6. Winstanley, P.A. (1995) *Baillière's Clin. Infect. Dis.,* **2**, 293–308.
7. Miller, K.D., Lobel, H.O. and Satriale, R.F. (1986) *Am. J. Trop. Med. Hyg.,* **35**, 451–458.
8. Inselburg, J. and Zhang, R.D. (1988) *Am. J. Trop. Med. Hyg.,* **39**, 328–336.
9. Zolg, J.W., Plitt, J.R., Chen, G.X. and Palmer, S. (1989) *Mol. Biochem. Parasitol.,* **36**, 253–262.
10. Chen, G.X., Mueller, C., Wendlinger, M. and Zolg, J.W. (1987) *Mol. Pharmacol.,* **31**, 430–435.
11. Sirawaraporn, W., Sirawaraporn, R., Cowman, A.F., Yuthavong, Y. and Santi, D.V. (1990) *Biochemistry,* **29**, 10 779–10 785.
12. Peterson, D.S., Milhous, W.K. and Wellems, T.E. (1990) *Proc. Natl Acad. Sci. USA,* **87**, 3018–3022.
13. Brooks, D.R., Wang, P. and Read, M. (1994) *Eur. J. Biochem.,* **224**, 397–405.
14. Slater, A.F. and Cerami, A. (1992) *Nature,* **355**, 108–109.
15. Foley, M., Deady, L.W., Ng, K., Cowman, A.F. and Tilley, L. (1994) *J. Biol. Chem.,* **269**, 6955–6961.
16. Bray, P.G., Howells, R.E. and Ward, S.A. (1992) *Biochem. Pharmacol.,* **43**, 1219–1227.
17. Krogstad, D.J., Schlesinger, P.H. and Gluzman, I.Y. (1989) *Prog. Clin. Biol. Res.,* **313**, 53–59.
18. Cowman, A.F., Karcz, S., Galatis, D. and Culvenor, J.G. (1991) *J. Cell Biol.,* **113**, 1033–1042.
19. van Es, H., Karcz, S., Chu, F., Cowman, A.F., Gros, P. and Schürr, E. (1994) *Mol. Cell. Biol.,* **14**, 2419–2428.
20. Barnes, D.A., Foote, S.J., Galatis, D., Kemp, D.J. and Cowman, A.F. (1992) *EMBO J.,* **11**, 3067–3075.
21. Krogstad, D.J., Gluzman, I.Y., Herwaldt, B.L., Schlesinger, P.H. and Wellems, T.E. (1992) *Biochem. Pharmacol.,* **43**, 57–62.

22. Cowman, A.F., Galatis, D. and Thompson, J.K. (1994) *Proc. Natl Acad. Sci. USA*, **91**, 1143–1147.
23. Wellems, T.E., Walker, J.A. and Panton, L.J. (1991) *Proc. Natl Acad. Sci. USA*, **88**, 3382–3386.
24. Bradley, D. and Warhurst, D.C. (1995) *Br. Med. J.*, **310**, 709–714.
25. Watkins, W.M., Woodrow, C. and Marsh, K. (1991) *Am. J. Trop. Med. Hyg.*, **49**, 106–112.
26. Watkins, W.M., Winstanley, P.A. and Murphy, S.A. (1995) *Br. J. Clin. Pharmacol.*, **39**, 283–289.
27. Nosten, F., ter Kuile, F.O. and Luxemburger, C. (1993) *Lancet*, **341**, 1054–1056.
28. Peters, W., Robinson, B.L. and Milhous, W.K. (1993) *Ann. Trop. Med. Parasitol.*, **87**, 547–552.
29. Hien, T.T. and White, N.J. (1993) *Lancet*, **341**, 603–608.
30. Meshnick, S.R. (1994) *Trans. R. Soc. Trop. Med. Hyg.*, **88** (Suppl. 1), 31–32.
31. White, N.J. (1995) *Baillière's Clin. Infect. Dis.*, **2**, 309–330.
32. Brewer, T.G., Peggins, J.O. and Grate, S.J. (1994) *Trans. R. Soc. Trop. Med. Hyg.*, **88** (Suppl. 1), 33–36.
33. Fry, M. and Pudney, M. (1992) *Biochem. Pharmacol.*, **43**, 1545–1553.
34. Vaidya, A.B., Lashgari, M.S., Pologe, L.G. and Morrisey, J. (1993) *Mol. Biochem. Parasitol.*, **58**, 33–42.
35. O'Donoghue, P.J. (1995) *Int. J. Parasitol.*, **25**, 139–195.
36. Georgiev, V. St (1994) *Drugs*, **48**, 179–188.
37. Ghoda, L., Phillips, M.A., Bass, K.E., Wang, C.C. and Coffino, P. (1990) *J. Biol. Chem.*, **265**, 11 823–11 826.
38. Fairlamb, A.H., Henderson, G.B. and Cerami, A. (1989) *Proc. Natl Acad. Sci. USA*, **86**, 2607–2611.
39. Pépin, J. and Milord, F. (1994) *Adv. Parasitol.*, **33**, 1–47.
40. Moreno, S.N.J., Mason, R.P. and Docampo, R. (1984) *J. Biol. Chem.*, **259**, 8252–8259.
41. Lloyd, D. and Pedersen, J.Z. (1985) *J. Gen. Microbiol.*, **131**, 87–92.
42. Ellis, J.E., Wingfield, J.M., Cole, D., Boreham, P.F.L. and Lloyd, D. (1993) *Int. J. Parasitol.*, **23**, 35–39.

Appendix A. Glossary

Allergic response: an antigen-induced response that is damaging to the host tissues. The host is hypersensitive to the antigen (allergen). Immediate hypersensitivity is mediated by antibody, but delayed hypersensitivity is cell-mediated.

Aminoglycoside: a family of basic antibiotics typically comprising three carbohydrate residues substituted with positively charged amino and guanido groups.

Anaphylactic response: IgE antibodies on the surface of mast cells and basophils interact with the IgE causing the cells to degranulate, releasing the contents of their secretory granules including histamine and serotonin.

Anti-metabolite: a substance that inhibits the metabolism of a substrate that often has a close structural similarity. The anti-metabolite inhibits an enzyme for which the target metabolite is a substrate.

Antiseptic: a preparation that may be used on the skin to reduce the bacterial count by killing or inhibiting the bacterial flora.

Aseptic: descriptive of a state of freedom from bacteria, or a technique or procedure designed to prevent the entry of bacteria.

Bactericidal drug: a compound that kills bacteria.

Bacteriostatic drug: a compound that does not kill, but prevents the growth and proliferation of bacteria.

Biliary excretion: secretion by the liver into the bile.

Chemotherapy: the use of natural, but especially synthetic, drugs to prevent or to treat infections or tumors.

Disinfectant: a solution of chemicals that kills bacteria (although not necessarily spores).

DNA polymerase: an enzyme that synthesizes DNA from the four deoxynucleotide triphosphates.

Enterohepatic circulation: a cycle of secretion by the liver into the bile, reabsorption from the intestine and return to the liver.

Hypersensitivity: *see* Allergic response.

Initiation (of protein synthesis): the assembly of the ribosomal subunits together with messenger RNA and formyl-methionyl tRNA to begin the synthesis of a new polypeptide chain.

Isobologram: a graph in which the concentrations of two drugs are plotted with a line showing all the combinations of the two that have the same effect.

Kernicterus: a jaundice of the newborn infant involving the deposition of bile pigments in the nervous tissues.

K_i: the inhibitor constant on terms of concentration of inhibitor. The lower value of the constant the more effective the inhibitor.

K_m: the Michaelis constant; the concentration of substrate that produces half the maximal velocity (V_{max}).

β-Lactam: compound containing a four membered ring comprising three carbon and one nitrogen atom. Adjacent to the nitrogen atom is a carbonyl group.

Lysis: the breakdown of cells due to damage to the cell walls followed by bursting of the membrane. It is marked by loss of cell contents from ions and metabolites to macromolecules.

Lysozyme: an enzyme found in tissues and body fluids that causes the lysis of bacteria. The enzyme hydrolyzes the peptidoglycan of the cell wall allowing the cell to burst under its own internal osmotic pressure.

Macrolide: a family of antibiotics comprising large lactone rings (12–16 atoms) with carbohydrate side chains.

Messenger RNA (mRNA): a sequence of RNA that can be transcribed at the ribosome to form one or more polypeptides.

Microsome: a vesicle of membrane and ribosomes that is produced by tissue homogenization and centrifugation. Microsomes are used to measure protein synthesis and the metabolism of drugs by inducible enzymes.

Mucopeptides: *see* peptidoglycan.

Mutation: change in the base sequence of a gene. Mutations may have an effect that is neglible (silent mutations) to profound, often causing inactivation of the gene product.

Ototoxicity: toxic effects on the auditory nerve.

Park nucleotides: compounds derived from precursors of peptidoglycan synthesis. These compounds accumulate in cells exposed to drugs that inhibit the synthesis of peptidoglycan.

Penicillin amidase: a fungal enzyme that is used to hydrolytically remove the side chains from penicillins produced by fermentation (like penicillin G). Hydrolysis forms 6-amino-penicillanic acid (the penicillin nucleus) which can then be acylated chemically to form a range of semi-synthetic penicillins.

Penicillinase: an enzyme that hydrolyses the β-lactam ring of penicillin, so destroying the antibacterial activity of the drug.

Penicilloic acid: the inactive product of hydrolysis of β-lactams by penicillinase.

Peptidoglycan: a lattice macromolecule that forms the mechanically strong layer of the bacterial cell wall. Comprises peptide cross-links between glycan chains.

Peptidyl transferase: an enzyme, a constituent of the larger ribosomal subunit, that makes the new peptide bonds in a nascent polypeptide.

Plasma protein binding: the property in drugs of binding to the constituent proteins of the plasma rather than being in free solution.

Plasmid: a small self-replicating DNA molecule, independent of the bacterial chromosome, that may encode several genes. Plasmids may enter bacterial cells (transformation) and confer the properties specified by their genes on the host cells.

Post-antibiotic effect: the inhibitory effect on surviving bacteria after exposure to an antibiotic.

Resistance: the property of a microorganism being unaffected by an inhibitor.

Ribosomes: particles upon which proteins are synthesized. Comprise large and small subunits both of which contain specific ribosomal proteins and RNA molecules.

RNA polymerase: an enzyme that synthesizes RNA from the four nucleoside triphosphates.

Super-infection: an unusual proliferation of some bacteria, not usually pathogenic, due to the suppression of others by antibiotics.

Therapeutic index: the ratio of inhibition towards the infectious agent and the infected host. A measure of the specificity of antimicrobial drugs.

Undecaprenyl lipids: bacterial lipids comprising eleven isoprene units that act as carriers for water-soluble carbohydrates, such as the peptidoglycan precursors, in the hydrophobic environment of the membrane.

Appendix B. Further reading

Andriole, V.T. (1988) *The Quinolones*. Academic Press, London.

Barrandell, L.B. and Fitton, A. (1995) Artesunate. A review of its pharmacology and therapeutic efficacy in the treatment of malaria. *Drugs*, **50**, 714–741.

Bentley, P.H. and Ponsford, R. (1993) *Recent Advances in the Chemistry of Antiinfective Agents*. Royal Society of Chemistry, London.

Ciba Foundation (1994) *Antimicrobial Peptides*, Ciba Foundation Symposium 186. John Wiley and Sons, Chichester.

Conte, J.E and Barriere, S.L. (1992) *Manual of Antibiotics and Infectious Diseases*. Lea and Febiger, Philadelphia.

Coulson, C.J. (1994) *Molecular Mechanisms of Drug Action*. Taylor and Francis, London.

Cozzarelli, N.R. and Wang, J.C. (1990) *DNA Topology and its Biological Effects*. Cold Spring Harbor Press, New York.

Croft, S.L. (1995) Antiprotozoal drugs: some echoes, some shadows. In *Fifty Years of Antimicrobials: Past Perspectives and Future Trends, 53rd Symposium of the Society for General Microbiology*. Cambridge University Press, Cambridge.

Crumplin, G.C. (1990) *The 4-Quinolones: Antibacterial Agents* in Vitro. Springer Verlag, London.

Duerden, B.I., Reid, T.M.S. and Jewsbury, J.M. (1993) *Microbial and Parasitic Infection*. Edward Arnold, London.

Edwards, D.I. (1990) in *Comprehensive Medicinal Chemistry*, Vol. 2 (C.Hansch, P.G. Sammes and J.B. Taylor, eds). Pergamon, Oxford, p. 725.

Franklin, T.J. and Snow, G.A. (1989) *Biochemistry of Antimicrobial Action*. Chapman and Hall, London.

Gale, E.F., Cundliffe E., Reynolds, P.E., Richmond, M.H. and Waring, M.J. (1981) *The Molecular Basis of Antibiotic Action*, 2nd edn. John Wiley and Sons, London.

Garrod, L.P., Lambert, H.P., O'Grady, F. and Waterworth, P. (1981) *Antibiotics and Chemotherapy*, 5th edn. Churchill Livingstone, Edinburgh.

Gilman, A.G., Rall, T.W., Nies, A.S. and Taylor, P. (1990) *The Pharmacological Basis of Therapeutics*, 8th edn. Pergamon, New York.

Hammond, S.M. and Lambert, P.A. (1978) *Antibiotics and Antimicrobial Action*. Edward Arnold, London.Harold, F.M. (1970) *Adv. Microb. Physiol.*, **4**, 45.

Herzberg, R.P. (1990) in *Comprehensive Medicinal Chemistry*, Vol. 2 (C. Hansch, P.G. Sammes and J.B. Taylor, eds). Pergamon, Oxford, p. 753.

Hooper, D.C. and Wolfson, J.S. (1993) *Quinolone Antimicrobial Agents.* American Society for Microbiology, Washington, D.C.

Howe-Grant, M. (1993) *Chemotherapeutics and Disease Control.* Wiley, New York.

Hugo W.B. (1967) *J. Appl. Bacteriol.*, **30**, 17.

Kleinkauf, H. and von Dohren, H. (1990) *Biochemistry of Peptide Antibiotics.* De Gruyter, Berlin.

Lancini, G., Parenti, F. and Gall, G.G. (1995) *Antibiotics: A Multidisciplinary Approach.* Plenum Press, New York.

Lorian, V. (1991) *Antibiotics in Laboratory Medicine.* Williams and Wilkins, Baltimore.

Mandell, G.L. and Sande, M.E. (1990) in *The Pharmacological Basis of Therapeutics*, 8th edn. (Goodman and Gilman, eds). Pergamon, Oxford, pp. 1047–1057.

McCormack, J.J. (1990) in *Comprehensive Medicinal Chemistry*, Vol. 2 (C. Hansch, P.G. Sammes, and J.B. Taylor, eds). Pergamon, Oxford, pp. 271–298.

Nikaido, H. (1994) *Science*, **264**, 382.

Pratt, W.B. and Fekety, R. (1986) *The Antimicrobial Drugs.* Oxford University Press, Oxford, pp. 229–251.

Rang, H.P., Dale, M.M. and Ritter, J.M. (1995) *Pharmacology.* Churchill-Livingstone, Edinburgh.

Reese, R.E. and Betts, R.F. (1993) *Handbook of Antibiotics.* Little Brown, Boston.

Reid, J.L., Rubin, P.C. and Whiting, B. (1992) *Lecture Notes on Clinical Pharmacology.* Blackwell Science Ltd., Oxford.

Rogers, H.J., Perkins, H.R. and Ward, J.B. (1980) *Microbial Cell Wall and Membranes.* Chapman and Hall, London.

Russell, A.D. and Chopra, I. (1990) *Understanding Antibacterial Action and Resistance.* Ellis-Horwood, Chichester.

Sammes, P.G. (1990) in *Comprehensive Medicinal Chemistry*, Vol. 2 (C. Hansch, P.G. Sammes, and J.B. Taylor, eds). Pergamon, Oxford, pp. 255–270.

Simon, C., Stille, W. and Wilkinson, P.J. (1993) *Antibiotic Therapy in Clinical Practice.* Schattauer, Stuttgart.

Siporin, C., Heifetz, C.L. and Domagala, J.M. (1990) *The New Generation of Quinolones.* Marcel Dekker, New York.

Sirotnak, F.M. *et al.* (1984) in *Folate Antagonists as Therapeutic Agents*, Vols 1 and 2. Academic Press, New York.

Sutcliffe, J.A. and Georgopapadakou, N.H. (1992) *Emerging Targets in Antibacterial and Antifungal Chemotherapy.* Chapman and Hall, New York.

Wakelin, L.P.G. and Waring, M.J. (1990) *Comprehensive Medicinal Chemistry*, Vol. 2 (C. Hansch, P.G. Sammes, and J.B. Taylor, eds). Pergamon, Oxford, pp. 703.

Winstanley, P.A. (1995) Anti-malarial chemotherapy. *Baillière's Clin. Infect. Dis.*, **2**, 293–308.

Index

ORDERING DETAILS

Main address for orders

BIOS Scientific Publishers Ltd
9 Newtec Place, Magdalen Road,
Oxford OX4 1RE, UK
Tel: +44 1865 726286
Fax: +44 1865 246823

Australia and New Zealand
DA Information Services
648 Whitehorse Road, Mitcham, Victoria 3132, Australia
Tel: (03) 9210 7777
Fax: (03) 9210 7788

India
Viva Books Private Ltd
4325/3 Ansari Road, Daryaganj, New Delhi 110 002, India
Tel: 11 3283121
Fax: 11 3267224

Singapore and South East Asia
(Brunei, Hong Kong, Indonesia, Korea, Malaysia, the Philippines,
Singapore, Taiwan, and Thailand)
Toppan Company (S) PTE Ltd
38 Liu Fang Road, Jurong, Singapore 2262
Tel: (265) 6666
Fax: (261) 7875

USA and Canada
BIOS Scientific Publishers
PO Box 605, Herndon, VA 22070, USA
Tel: (703) 435 7064
Fax: (703) 689 0660

Payment can be made by cheque or credit card (Visa/Mastercard, quoting number
and expiry date). Alternatively, a *pro forma* invoice can be sent.

Prepaid orders must include £2.50/US$5.00 to cover postage and packing
(two or more books sent post free)